Sustainable Development in the Digital Era

Collaborative Management

Leonilde Varela
University of Minho
Department of Production Engineering
Guimarães, Portugal

Goran Putnik
University of Minho
Mechanical Engineering Department
Guimarães, Portugal

Vijaya Kumar Manupati
Assistant Professor, Operations and Supply Chain Division, Indian Institute of Management Mumbai, India

CRC Press
Taylor & Francis Group
Boca Raton London New York

CRC Press is an imprint of the
Taylor & Francis Group, an **informa** business
A SCIENCE PUBLISHERS BOOK

First edition published 2025
by CRC Press
2385 NW Executive Center Drive, Suite 320, Boca Raton FL 33431

and by CRC Press
4 Park Square, Milton Park, Abingdon, Oxon, OX14 4RN

CRC Press is an imprint of Taylor & Francis Group, LLC

Library of Congress Cataloging-in-Publication Data (applied for)

ISBN: 978-1-032-64454-7 (hbk)
ISBN: 978-1-032-67370-7 (pbk)
ISBN: 978-1-032-67365-3 (ebk)

DOI: 10.1201/9781032673653

Typeset in Palatino Linotype
by Prime Publishing Services

We would like to thank the following for their valuable assistance:

Justyna Trojanowska (I - Chapter 4)
Gaspar Vieira and Cátia Alves (II - Chapter 1)
Veerababu Ramakurthi (II - Chapter 2)
Anupama Tiwari and José Machado (II - Chapter 3)
Ramesh Kumara, Anupama Tiwari, Om Prakash Kumawat and Koona Samatha (II - Chapter 4)
Madhulika Saxena, Niharika Patel and Veerababu Ramakurthi (II - Chapter 5)

Summary

Currently, in the digital age, collaborative manufacturing and management is fundamental to permit and promote a sustainable development of companies, either traditional or extended and networked organizations, including cyber physical systems.

This book aims at putting forward higher-quality technical-scientific content about fundamental methodologies, models, methods, tools, and platforms about collaborative engineering, to support manufacturing and management processes and practices, aligned with the current requirements underlying Industry 4.0, and Society 5.0 principles and aims. To this end, this book's focus is on the exploration and application of collaborative management paradigms about dynamic, distributed, integrated, intelligent, predictive, parallel, and real-time based approaches and tools to enable collaborating entities, including suppliers, business partners and other stakeholders, to develop projects and solve problems that are becoming increasingly more complex and challenging currently. Such collaborative processes and practices require companies and underlying stakeholders to be connected, and to further communicate, and share data, problems, and expertise, and other kind of resources, along with concerns, difficulties, and challenges, requiring co-learning, and the co-creation of knowledge, processes, methods, and systems to interactively support projects and problems solving.

In this book advanced ideas and works compendium, supported by case studies will be provided, to assist scientists, practitioners, and students in high standard manufacturing management processes and practices, to properly handle their daily base problems and challenges, with a special focus on the use of recent paradigms and tools to support manufacturing management decision making, through innovative methodologies and approaches for permitting researchers to learn, develop further work, and become advanced practitioners and promoters of collaborative management.

Preface

Collaboration between and within companies, along with suppliers, customers, and stakeholders is crucial in the current digital era to enable a global sustainable development of organizations, varying from more traditional to extended and virtual ones. Recent developments underlying the Industry 4.0 concept, in Europe and further the Society 5.0 one, in Japan, enable and promote Collaboration, by putting forward technology that permits connection, communication, sharing, co-learning, and co-creation of knowledge, methods, and tools between those entities. Collaborative Manufacturing and Management processes and practices, is thus currently possible and easier to be accomplished through the use of advanced technology that permits putting into use a set of fundamental underlying paradigms, about dynamic, distributed, integrated, intelligent, predictive, parallel, and real-time based approaches, systems, and platforms, which are of prime relevance currently for managing either more traditional or more advanced, smart and cyber physical production systems.

Scope

With this book, the editors aim at compiling a higher-quality content about fundamental methodologies, methods, models, tools, and platforms that enable putting into practice collaborative engineering principles, which is of prime importance currently for worldwide researchers, managers and stakeholders, to provide a further understanding about the current need for using such collaborative engineering principles, namely in the scope of manufacturing management. It is thus expected that the set of advanced ideas and works compendium, supported by case studies, will assist scientists and practitioners in high standard manufacturing management processes and practices, to properly handle their problems and challenges on a daily basis, with a special focus in the new areas of manufacturing management paradigms, and to create a basic knowledge resource for young researchers through which they can learn and develop further towards becoming advanced practitioners and promoters of the mentioned subject.

Target Audience

The current perspectives of Collaborative manufacturing management paradigms, approaches, and systems hold important implications for current practices and understanding these concepts and processes for further implications consists of an emergent need. Moreover, considering environmental and/or social performance, and economic performance integrated with manufacturing system performances need to be further understood. This book aims at integrating significant knowledge in the focused domain, and to become very helpful for Graduate, and Postgraduate students as well, along with advanced managers, decision-makers, practitioners, and researchers by elucidating about its practical implications.

Contents

Part I

Fundamentals of Collaboration and Collaborative Manufacturing Management

1

Main Collaboration and Collaborative Engineering Concepts

Collaboration and Collaborative engineering are not new topics but have taken on new importance today, in the context of Industry 4.0 or I4.0 for short. There are other concepts, which, although well known, are different but often mismatched with collaboration. Therefore, the main objective of this chapter is to present a concept of collaborative engineering, together with the underlying sub-concepts, supported by an extensive systematic literature review carried out. A critical analysis and discussion is also presented, based on the main insights brought through the literature review. The fundamental importance of collaborative learning, along with the human role, in the current I4.0 is also discussed in this chapter, based on a proposed collaboration framework.

Introduction

The term collaboration is frequently used for expressing some kind of joint process or practice to be carried out, for instance, in manufacturing management context (Mowery, 1989; McClellan, 2002; Huang, 2002; Sluga, Bilbao et al., 2004; Sluga et al., 2005; Nof, 2006; Chituc et al., 2007; Lin et al., 2008; Wang, 2008a,b; Zaletelj et al., 2008; Camarinha-Matos et al., 2009a,b; Cherubini et al., 2016; Arrais-Castro et al., 2018; Feng and Huang, 2018; Yang et al., 2017), or in some other engineering or manufacturing decision-making processes (Alavi et al., 1995; Barnes et al., 2006; Raike et al., 2013; Karp and Pardo, 2017; Blackburn et al., 2020; Moreno-Guerrero, et al., 2020; Sousa et al., 2021).

One of such frequent uses of the collaboration occurs, in fact, in the Engineering field, and it is referred to as collaborative engineering (CollE) (Putnik et al., 2021a,b), which is often confused with the concurrent engineering (CE) concept (Putnik et al., 2021a,b), whereby assuming that a main common goal does exist, but which in fact does not need to be the case, either in a more tangible or manufacturing practices scenario (Jiang and Zhang, 2002; Shyamsundar and Gadh, 2002; Simatupang and Sridharan, 2002; Deek et al., 2003; Colombo et al., 2004, 2005; Leitão et al., 2005; Nof, 2006; Li et al., 2007; Zaletelj et al., 2008; Ma et al., 2008; Camarinha-Matos et al., 2009; Belkadi et al., 2020) or in a more intangible or management context, namely, regarding some collaborative information or knowledge processing or some other kind of decision-making process (Mitchell & Singh, 1996; Fröhlich et al., 1997; Svendsen, 1998; Majchrzak et al., 2000; Aviv, 2001; Bititci, et al., 2003; Ulaga, 2003; Ming et al., 2005; Heckscher and Heckscher, 2007; Ramstad, 2008; Brown et al., 2016; Arrais-Castro et al., 2015, 2018; Sousa et al., 2021).

The main difference between the two concepts (CollE and CE), as referred in (Putnik et al., 2021a,b), is that in CollE there may or may not exist a common goal, but, instead, it implies the existence of a common understanding, learning or co-learning (van Eijnatten et al., 2004; Putnik et al., 2021a,b).

Moreover, the existence of a common understanding as a necessary condition for collaboration within a manufacturing environment pre-assumes the existence of a common communication standard, a common language between the elements of the community. Such communication standards are the real necessary condition for the very emergence of collaborative networks, for instance, IIoT communication protocols (Lin et al., 2015; Li et al., 2017).

Besides, regarding more practical application scenarios occurring, for instance, in a manufacturing environment, it frequently implies sharing some kind of material or manufacturing resource, such as a production tool or machine (Ferreira et al., 2022). It may also just imply sharing some piece of information, knowledge, or experience (Ferreira et al., 2022).

In any case, according to this perspective, in a collaborative procedure or practice, to be called as such, some kind of mutual or joint learning process has to occur (Putnik et al., 2021a,b). In this regard, and depending on the type of collaboration inherent to a specific application scenario, different kinds of approaches, means and tools, or systems can be explored for establishing diverse types of collaboration processes or practices, which include Human-Human, Human-Machine, and Machine-Machine types of collaboration (Ferreira et al., 2022).

In this chapter, the collaboration concept will be analysed through the lenses of existing literature, in order to properly support the main ideas underlying a proposed collaboration concept.

Collaboration is a word that comes from the word 'colaborare', and from its origin, it means to express the action of working or operating with some other entity (e.g., someone) or anything, to produce or create something (https://en.wikipedia.org/wiki/Collaboration).

In the literature can be identified two distinct philosophical currents about the term collaboration, starting from the goal issue. One that defends the existence of a common goal or objective, and another one for which this does not have to be verified.

Collaboration is defined as: "social skills, relationships, practices, and technology services that improve how people work jointly and substantially together (sharing responsibility and risk), to communicate needs, coordinate activities, share information, exchange know-how, build community or achieve a common (team) objective (typically related to a process or project) within or across organi sational boundaries' (http://mikeg.typepad.com/perceptions/2004/05/defining_collab.html)

In Camarinha-Matos and Afsarmanesh (2006) and Abreu and Camarinha-Matos (2008), the authors further state that collaboration is, in fact, not just related to sharing or distributing data, and information, but also with sharing knowledge, benefit, profit, skills, competences, along with costs, dependencies, difficulties, and even risks, between two or more entities.

In Lai (2011), collaboration is defined as being a "mutual engagement of participants in a coordinated effort to solve a problem together". Moreover, the authors refer that "collaborative interactions are characteris ed by shared goals, symmetry of structure, and a high degree of negotiation, interactivity, and interdependence…".

These considerations are, in fact, subjacent to the very closely related concept of concurrent engineering (CE) which is already very well established in the literature (Putnik and Putnik, 2019).

Summarising, the origin of the collaboration concept enables us to highlight that:

- Collaboration is a word that originates from the word '**colaborare**' and since its origin it aims to express the **"action of working or operating with some other entity…"** (for example, one person with another, to produce or create something together (Schrage, 1990) in *Shared Minds*.

- Collaboration is also defined as: **"social skills, relationships, practices, and technological services that improve the way people work together, to communicate needs, coordinate activities, share information, exchange know-how, build a community, or achieve a common goal"** (of a team, typically related to a process or project) within a given organis ation or between organis ations (Gotta, 2008).
- In the literature, **two main philosophical currents are identified** about the term collaboration, one regarding the **existence of a common goal** (Camarinha-Matos and Afsarmanesh, 2004, 2006, 2007; Gotta, 2008), and another one about the **existence of a common understanding** (Putnik et al., 2021a,b; Varela et al., 2022; Varela et al., 2022,a,b,c).
- Abreu, and Camarinha-Matos, in 2008 emphasis ed that **"collaboration is not only related to the sharing of data and information, but also to the sharing of knowledge, benefits, profits, capabilities, skills, and also with the sharing of costs, dependencies, difficulties, and even risks between two or more entities, …"**.
- Lai (2011) defined collaboration as a **"mutual engagement of participants in a coordinated effort to solve a problem together**, with collaborative interactions **characteris ed by shared goals and a high degree of negotiation, interactivity, and interdependence …"**.
- Putnik and Putnik (2019) mentioned some **differences between more or less closely related engineering strategies**, namely, referring to **the typical existence of conflicts in concurrent engineering, requiring negotiation processes, while highlighting the importance of the dialogue in collaboration.**

In order to properly define collaboration, a systematic literature review (SLR) was performed, with an additional objective to identify main collaboration sub-concepts, and types of collaboration, further correlating them, and extracting information about main clusters of terms being used, regarding the Collaboration paradigm, including main methodologies, approaches or methods, and key terms, that are going to be summaris ed in this chapter.

Proposed Collaboration Concept

In this section, a CollEng-M&M conceptual model, along with its main underlying components or sub-concepts will be put forward, based on the authors' own knowledge and experience, underlying main research activity, about core findings related to collaboration, along with some main findings from the literature, that were further deeply analysed, through the results of the SLR conducted, to properly support the proposed conceptual model, and by briefly describing the main outcomes reached (Varela et al., 2022).

The proposed CollEng-M&M conceptual model consists of six main pillars that address some considered main conditions, structured as shown in Figure 1.

Proposed Collaboration Conceptual Model

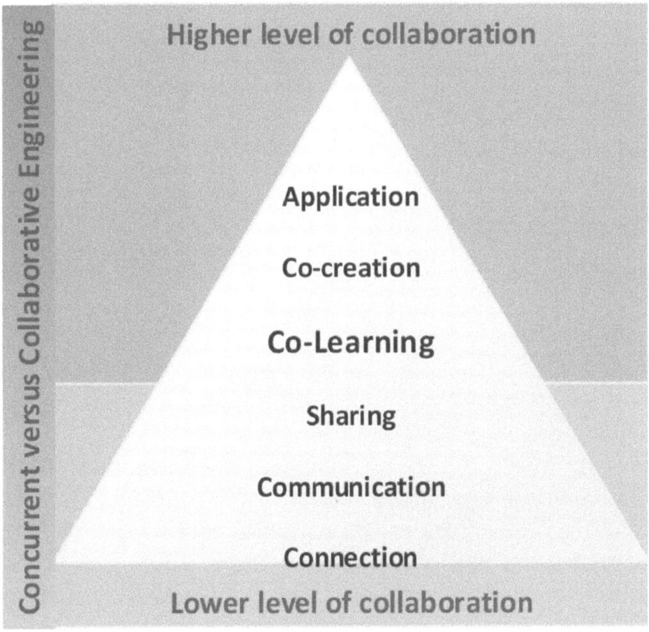

Figure 1. Collaborative engineering and management conceptual model (adapted from: Varela et al., 2022).

This conceptual model includes two main areas of sub-concepts:

A lower level of collaboration—with the inclusion of connection, communication, and sharing of information, know-how, technology, means, approaches, tools or platforms, therefore tangible and/or intangible assets, namely, **through some kind of coordination or cooperation mechanism**, but **without effective (co)learning**, therefore, just satisfying a basic level or form of collaboration, that is characterised as a partial collaboration.

A higher level of collaboration—with the inclusion of (co)learning through appropriate technology, means, approaches, tools, or platforms that allow, and envision some type of co-creation, centred or not on human intervention, namely, in manufacturing and/or management processes and practices.

Collaboration Sub-concepts

Connection consists of some kind of physical or logical link between two or more things or entities. In case of more than two entities being connected, it is also usually referred as an 'Interconnection' (https://en.wikipedia.org/wiki/Connection).

Main connection/connectivity types

- Networked supply chains
- Connected companies
- Connected machines
- Connected products
- Connected workers
- Connected services

Communication is the imparting or exchanging or transmission of information. Thus, the concept or state of exchanging data or information between entities.

For instance, a message is an example of data or information transferred in an act of communication (https://en.wikipedia.org/wiki/Communication).

Thus, communication is an instance of information transfer, and a required element for enabling a conversation or discourse.

Another example from the educational context can be referred to as being the professors' communications of lively discussion or via email. So a communication is carried out through a passageway or opening between two or more locations, the connections.

Connection and communication together enable to establish networks and underlying interactions, usually called network communications, through which information transmission or sharing occurs.

Network communications based on wireless and internet technologies (Internet of Things, IoT), along with the use of a widened set of communication technology, tools, devices, and means, e.g., sensors, actuators, Radio Frequency Identification (RFID), along with smart objects, among other technology, serves to link machines, work products, systems and people, within a manufacturing plant, intra company or inter companies, through a more or less extended network of stakeholders, which may include, e.g., suppliers, distributors, business partners, and customers communicating worldwide.

Schrage (1990) mentions that it is noticeable that a lack of structures allow people to express their competences creatively. Moreover, he says that merely teamwork does not mean authentic collaborative work, and that "shared spaces" like blackboards or brainstorming sessions enable it to pass from mere communication to true collaboration.

Main communication types

- Communication networks (through the cloud, ...)
- Several technologies (IoT, RFID, ...)
- Smart objects (devices, etc.)

...

Sharing is the joint use of a resource or space. It is also the process of dividing and distributing. In its narrow sense, it refers to joint or alternating use of inherently finite goods, 'sharing' can actually mean giving something as an outright gift: for example, to 'share' one's food really means to give some of it as a gift. Sharing is a basic component of human interaction, and is responsible for strengthening social ties and ensuring a person's well-being (https://en.wikipedia.org/wiki/Sharing).

In general, sharing consists on partaking or contributing with some kind of tangible or intangible asset, for instance:

- Sharing files, links, videos, data, and its processing, analysis and exploitation, either immediately on the factory floor, or in a broader sense, through the web or cloud.
- Sharing knowledge, competences, expertise, and skills.
- Sharing different kinds of resources, e.g., manufacturing resources, processors, machines, tools, etc.
- Sharing tasks, problems, costs, challenges, dependencies, risks, concerns, or difficulties.
- Sharing technology, techniques, software, benefits, innovations, time, and thinking.
- Sharing suppliers, business partners, products, materials, production systems, warehouses, transportation means and logistic systems, companies, and customers.

Main sharing types

- Data, information, knowledge
- Resources, machines
- Processes, tasks, warehouses, transportation means,

Learning consists of the acquisition of knowledge or skills through study, experience, or being taught (https://en.wikipedia.org/wiki/Learning), and an important requisite for this knowledge or skills acquisition is the existence of feedback among entities.

Eijnatten and Putnik (2004) distinguish between different kinds of learning sub-concepts, further based on other existing sources, as follows (Eijnatten and Putnik, 2004):

Collective co-learning: "The ability of the collective to learn from experiences drawn by individuals while working. It is one single phenomenon that consists of four abilities: relationics, correlation, internal model, and praxis (Backström, 2004)."

Collaborative learning or co-learning: "Learning that occurs as a result of interaction between peers in the completion of a common task (Noble, 2004)."

In (https://en.wikipedia.org/wiki/Learning), it is also described the "Organis ational learning" concept, as follows:

Organis ational learning (OL): "is defined as the way people jointly construct maps (Argyris and Schön, 1997) or exercise competence and enact qualifications in a network of interacting people (Jensen & Rasmussen, 2004). OL is about the learning process, and more specifically about the co-operative learning process (McHugh et al., 1998) in a specific socio-cultural context (Cullen, 1999). Moreover, OL introduces hierarchical levels of learning, i.e., single loop (correction of errors by using feedback), double loop (changing underlying norms and mechanisms), and triple loop (questioning essential principles, learning about learning), and also includes organisational processes. In a critical review, Tsang (1997) states that in OL, change cognition is a necessary condition (Tsang, 1997): "The cognitive aspect is generally concerned with knowledge, understanding, and insights." But according to Eijnatten and Putnik (2004), there is a split among definitions on whether a change in actual or potential behaviour is required, and by potential behavioural change, these authors assume that the lessons learnt by an organisation would have an impact upon its future behaviour." Eijnatten and Putnik (2004) further explored the concept of Learning Organisation as follows:

Learning Organi sation (LO): "An organisation, structure, process, or network "where people continually expand their capacity to create the results they truly desire, where new and expansive patterns of thinking are nurtured, where collective aspirations are set free, and where people are continually learning to see the whole together" (Senge, 1990). The five required disciplines or "component technologies" are: (1) Team learning, (2) Building shared vision, (3) Mental models, (4) Personal mastery, and (5) Systems thinking (Senge, 1990)."

Additionally, the LO concept is also seen or defined in Eijnatten and Putnik (2004) as: "An organisation, structure, process, or network 'which is capable of thriving in a world of interdependency and change' (Kofman and Senge, 1993)."

Further described in Eijnatten and Putnik (2004) as being: "A social system whose members have learned conscious communal processes for continually generating, retaining and leveraging individual and collective learning to improve performance of the organisational system in ways important to all stakeholders; and monitoring and improving performance (Drew and Smith, 1995)."

Besides in Eijnatten and Putnik (2004), the LO is also described as a: "Sum total of accumulated individual and collective learning (Hyland and Matlay, 1997)."

Moreover, Eijnatten and Putnik (2004) also mentioned the LO as: "An organis ation, structure, process, or network 'exhibiting directed changes at the macro level' (Jensen and Rasmussen, 2004)."

Further, in Eijnatten and Putnik (2004) is also mentioned that "according to Huysman (2000), an LO is 'a form of organisation that enables the learning of its members in such a way, that it creates positively valued outcomes, such as innovation, efficiency, better alignment with the environment and competitive advantage'." Also that: "According to Huysman (2000), an LO is an organisation capable of adapting, changing, developing, and transforming itself in response to needs, wishes, and aspirations of people both inside and out." And that: "Both structural and cultural organis ation learning mechanisms are important to create and maintain a LO (Pedler et al., 1991)".

Besides Eijnatten and Putnik (2004), Sun & Scott also state that: "Learning must transfer from individual(s) to collective(s) to organisational to inter-organis ational, and vice-versa, and 'must' result in changes in behaviour (Sun and Scott, 2003)".

Further, as a concluding remark about the learning process in Eijnatten & Putnik (2004) it is stated that:

"In the most general sense, learning may be described as an iterative process of activities, whereby new knowledge is produced through transformation of experience. Whenever knowledge is created as a result of individual experiences, i.e., walking through the cycle of planning, decision-making (DM), action, experience, and reflection, we speak of "individual learning" (IL) (Kolb, 1984). When it results from interaction between peers, we speak of "collective learning" (CL) (Backström, 2004) or "collaborative learning" (Noble, 2004). The outcomes of IL and CL may be used for either personal or communal purposes, such as the further development of their own company."

Based on these main ideas and definitions, it is possible to draw a set of learning means and outcomes as follows:

- Learning through shared experiences, and goals, regarding individual and collective learning approaches and practices;
- Learning manufacturing processes, and operations, in the context of H-H, H-M, and M-M collaboration;
- Learning M&M, and underlying DM processes, methods, and tools;
- Learning improved ways of interactions with worldwide companies' stakeholders, e.g., suppliers, business partners, and clients;
- Learning innovations and education methodologies, etc.;
- Learning everything needed or wanted by interacting with someone and/or through something (organisation, social network, etc.).
- Summarising, the previously expressed main ideas inherent to the learning concept, and summarising, as stated in (Eijnatten and Putnik, 2004): "learning may be described as an iterative process of activities, whereby new knowledge is produced ...". Moreover, as a "... LO is "a form of organisation" that enables the learning of its members in such a way, that it creates positively valued outcomes, ...", and this leads to a kind of 'natural linkage' of the learning concept to the next one defined in the proposed CollEng-M&M conceptual model, the co-creation.

Main co-learning types

Co-learning between humans

Co-learning between machines

Co-learning between humans and machines

Co-learning between organisations

Co-learning tasks execution (by humans or machines, ...)

Co-learning different subjects (in smaller or bigger communities, e.g., in extended or (Run on) networked environments)

Co-creation is a general concept that can be used to define a widened set of 'things' that can be created, which may be intangible, such as more or less simple thoughts, or an idea or some more complex piece of information or knowledge, by a set of two or more members or entities interacting through some means and kind of learning process. On the other hand, in the tangible case, co-creation can further arise through diverse kinds or interactions, based on the underlying learning process, depending on the concrete type of means and materials used among the two or more interacting entities.

In the Wikipedia, in the context of a business, it refers to "a product or service design process in which input from consumers plays a central role from beginning to end". Less specifically, the term is also used for "any way in which a business allows consumers to submit ideas, designs, or content''. This way, a firm will not run out of ideas regarding the design to be created and at the same time, it will further strengthen the business relationship between the firm and its customers. Another meaning is ''the creation of value by ordinary people, whether for a company or not' (https://en.wikipedia.org/wiki/Co-creation).

Co-creation was defined by Jansen and Pieters, in 2017, as "a transparent process of value creation in ongoing, productive collaboration with, and supported by all relevant parties, with end-users playing a central role".

The co-creation term is already a concept relatively in regular use, especially in marketing and some design practices (e.g., open design), and in these disciplines it refers to a joint design of a product by designer and customer and further extensible to a more or less enlarged set of participating product development members, working together as a collaborative team.

This frequently mentioned co-creation concept is thus relatively close to the 'traditional' Concurrent Engineering (CE) concept that also requires some close relation and practices between a set of members in a team, working to usually reach some common or concurrent goal or objective, and which typically implies some kind of negotiation process (Putnik and Putnik, 2019).

However, the semantics is quite different from theory to theory, from author to author, from "user"-group to "user"-group, from community to community, and frequently CE and CollEng are mixed up or undistinguished, being thus frequently used as similar concepts (Putnik and Putnik, 2019; Putnik et al., 2021b, 2021c).

Therefore, in this paper, the objective is to clarify the CollEng concept, as being different from the CE one, as in fact we consider that in CollEng there is no need to have a common goal, but, instead, a common learning, between the collaborating members.

Besides the co-design or open design, it is also possible to consider some other kind of co-creation, for instance, co-operation through human-robot collaboration, and further, based on any other possible co-creation type, such as: co-learning, co-decision, co-management, co-maintenance, co-transportation, and co- or open-innovation, among others.

One such specific term arising from the general co-creation one, which is frequently used is co-work or co-working, mentioned in (Petrillo et al., 2018), and considered to be crucial up from the I4.0 and abroad to the next, the fifth industrial revolution (Industry 5.0 or I5.0, for short) (Nahavandi, 2019).

Another closely related term to co-working is cooperating or cooperation. As stated in (Bechtold and Lauenstein, 2014), Japan begins to talk about this fifth industrial revolution, which will be marked by the cooperation between man and machine.

The authors in (Petrillo et al., 2018) claim about this importance of co-working in the context of an expected significant increase in the complexity of production environments, and corresponding problems to be solved, leading to a growing need for further interactions, for instance between humans and machines, which is the so-called H-M collaboration (Varela et al., 2022; Manupati et al., 2022).

Finally, in the Engineering field, the application or practical implementation of the 'co-creation' in some specific domain is also of utmost importance.

Main co-creation types:

Co or open design or project

Co-processing (data, information, knowledge, ...)

Co-analysis (data, information, knowledge, ...)

Co-decision

Co-working

Co-transportation

Co-maintaining

Application is defined in (https://en.wiktionary.org/wiki/application) as being the act of applying as a means; the employment of means to accomplish an end; a specific use. Also, the act of directing or referring something to a particular case, e.g., the application of a theory to a set of data, etc.

The application level is thus considered a key issue in the proposed CollE-M&M concept, as in the engineering context it is naturally assumed and required to exist as some kind of application or implementation.

According to the kind of co-creation underlying the earlier stage or level in our proposed CollE model, the application can thus vary in a corresponding widened range of alternative scenarios, deriving from the underlying specific co-creation type.

For example:

Through co-learning or co-innovation, some new co-created concept of knowledge can be synthesised or formalis ed and further applied in some specific application domain.

In the case of some kind of co-working, such as in a team of two or more people working together in some shared document, e.g., by using google docs, a final document will be jointly produced.

Also, in the co-design or open design application scenario, as a result or application will derive some new product design, through a so-called H-H and/or H-M collaboration.

In a similar way, through co-management or co-decision, some important conclusion or decision can be taken to be implemented.

Also through co-maintenance or co-transportation, some kind of task can be carried out by a group of collaborating people and/or jointly through some kind of means, tools, machines, or transportation device.

Further, in the case of some kind of co operation, or, for instance, in some H-M collaboration scenario, for instance, through human-robot collaboration, some kind of task will be accomplished jointly by a human and a robot, and in some context of M-M collaboration, two or more machines or robots can also cooperate to reach some specific objective or accomplish some kind of task together. Therefore, cooperation or mere coordination between two or more entities, namely between two or more machines, e.g., robots, is in fact just a lower-level of collaboration, as not implying co-learning and/or true co-creation, which is clearly the case in the M-M collaboration context, without human intervention.

Main application types

Co-creation of software, documents, …

Group decision-making in manufacturing management, …

Collaborative robotics, …

Therefore, we may conclude that collaboration does, in fact, further imply some kind of application for reaching the proposed full CollE-M&M concept, for instance, through co-design (open design), co-work, co-operation, co-maintaining, co-monitoring, co-visualizing, co-learning, co-thinking, co-data handling, co-analysing, co-interpretation, co-deciding, co-sensing, co-reasoning, etc., resulting in some kind of output, which, in the concrete engineering scope, will be referred as being some kind of application or implementation, either through a more tangible or intangible asset.

In accordance with the information previously exposed, it is now possible to present a collaboration definition.

Collaboration Definition

Collaboration can be defined as an **interaction between two or more entities** (e.g., people, machines or both combined) **connected and communicating** by using some means and methods, being supported and enhanced through the use of appropriate technology to enable **sharing any kind of tangible or intangible resource** (e.g., information, problem, means, machines, etc.), and **co-learning, based on openness and common understanding, to permit the emergence of contributions and reaching consensus, to co-create some tangible or intangible asset** (e.g., information, knowledge, method, product, tool, or system), aiming at **some application or implementation**, being considered **essential for**

reaching true innovation and **to promote a sustainable development of companies in the current digital age**.

Next, the main types of collaboration will be briefly described.

Types of Collaboration and Relation with Collaboration Levels

Collaboration can be applied in different forms for reaching diverse kinds or **types of collaboration:**

- Human-Human collaboration (H-H): Such as, co-work based on shared resources, e.g., google drive and docs, etc.
- Human-Machine (H-M) or Human-Resource (H-R) collaboration: For instance, based on the use of group DM approaches and methods, among other methodologies and approaches, e.g., based on AI approaches and methods, along with a varying kind of meta-heuristics, etc., namely for supporting joint decision- making processes; human-robot work or cooperation; other kinds of H-M collaboration, for example, for machine training, e.g., in supervised machine learning (ML) context, though the use of "the oracle", or based on other approaches, such as, based on game theory, among others; or through some kinds of H-M co-work, e.g., co-design, co-maintain, co-transportation, etc.
- M-M or R-R collaboration: E.g., through integrated automatic or autonomous processes (ex: use of multi-agents, blockchain and smart contracts, big data (BD) processing, namely based on AI approaches such as machine or deep learning algorithms, among others, for instance to enable chaos and complexity processing and analysis, or the application of game theory methods, etc.

Human-Human Collaboration (H-H) examples

- Group learning (Ex: a group of two or more people learning some subject together)
- Group decision-making (Ex: use of a multi-criteria group decision-making method to obtain a decision by a set of industrial managers)
- Group work (Ex: two or more people co-editing a document by using a tool from Google Docs, among others)

Human-Machine (H-M) or Human-Resource (H-R) Collaboration examples

- Human-robot co-work (Ex: Semi-automatic inspection of the surface finish quality of a product, e.g., of a mobile phone assembled in a shop floor through an interaction between a human operator and an integrated robotic vision system)
- Machine/algorithm training (Ex: Oracle providing positive and negative examples to a supervised machine learning algorithm)
- Human-Company collaboration, namely, through the figure of a kind of 'broker' in the context of virtual communities or networks of companies

Machine-Machine (M-M) or Resource-Resource (R-R) Collaboration examples

- Collaboration between machines/algorithms (Ex: a set of two or more algorithms processing a distributed algorithm in parallel to reach a joint final solution for a given problem)
- Collaboration between robots (Ex: group of two or more robots cooperating in the execution of some manufacturing operation)
- Collaboration between two or more companies or organisations, for instance in the context of producing some product or while providing some service in the context of a Virtual community, integrating a network of companies collaboration in the execution of product's tasks.

Relation between Collaboration Levels and Types

According to the proposed collaboration, conceptual model put forward, and previously described, in the Section 3, in this work the main categories of publications were organised considering a partial or incomplete use of the collaboration concept (without 'learning'), e.g., through IoT, cloud computing or manufacturing, augmented reality, mixed reality or digital twins (DT) based approaches, purely based on technology, and without explicit learning (co-learning) practices, even based on some other collaborative tools, such as, serious games, google docs, etc.). Although, in this cluster of publications, there is at least subjacent some connection, communication, resources sharing or common data handling, based on some kind of Information and Communication Technology (ICT) infrastructure and/ or cyber-physical system (CPS) (Romero et al., 2016; Stern and Becker,

Table 1. Main characteristics of lower and higher levels of collaboration clusters with and without human intervention (adapted from Varela et al. (2022)).

Collaboration Type Level	Not human-based (M-M)	Human-based (H-H, H-M)	Main Skills
2nd level (higher level): with learning, and, co-creation	Machine/deep learning; advanced automation; self-organis ation; self-parametrization, semantic web; neural networks; intelligent systems, patterns recognition, based on exponential technology, collaborative robots, AI, CPS, ...	Learning organis ation; supervised machine learning; semiotics; pragmatics; emergence; innovation; through co-x (co-design or open design, co-conceptualize, co-learn, co-decide, co-evolve, co-innovate, co-analyse, co-do, co-think, co-construct), ...	Hard and Soft skills: business/ organisational oriented: paradigms, models, methodologies, methods, systems, and platforms based on feedback, communication, dialogue, and, higher level approaches, through learning paradigms, means, and software, along with other technological support
1st level (lower level): connection, communication, and sharing, without co-learning	Digitali sation; integration, smart systems; smart objects, automatis ation, serviti sation, point-to-point, end-to-end communication, and processing...	Recommendation models and systems, DSS, human-machine interactions, human-robot, co-x: co-work, co-act, co-produce, co-operate, co-maintain, co-transport...	Mainly hard skills oriented -> models, architectures, methods, algorithms, devices, systems, and platforms based on data, knowledge, mainly transactional processes, means, and tools
Main focus	cooperation	collaboration	

2017; Shi et al., 2011; Emmanouilidis et al., 2019; Baheti and Gill, 2011; Bousdekis et al., 2020; Fantini et al., 2020), for reaching some kind of cooperation or joint/shared decision-making process, in some kind of practical application domain, in the industrial (manufacturing and/or management) context. Thus, it continues to be considered some kind of collaboration, of H-H, H-M, or M-M type.

In the presence of' learning', a more significant, higher, full or complete accomplishment of the proposed collaboration concept (with 'learning' or 'co-learning') is present, and which may be either "Not human centred" or "Human centre d"). In the case of being "Not human centred", examples such as pure M-M learning approaches, based on ML or on other kinds of procedures, for instance, based on Multi-agents' interactions, through

Multi-Agent Systems (MAS) may occur. In the case of the human presence being a key factor, this category is considered the most relevant one, in the scope of the proposed CollE concept, being H-H: B2.1), and H-M: B2.2) collaboration types considered, and which are marked by some kind of co-learning practice, and further applied in some kind of industrial (M&M) context, thus reaching the higher or complete level of collaboration.

Table 1 synthesises the main clusters of publications and underlying principal characteristics regarding cooperation and collaboration issues, in order to properly and clearly state the collaboration concept and subjacent contributions from the literature.

Final Discussion

According to the SLR conducted, some additional main ideas arose from the literature analysis and synthesis performed, and some main issues are briefly presented next.

In Li and Qiu (2006), the authors refer to research works and commercial systems that have been put forward to provide solutions for what they mention to be collaborative and distributed product development processes, by further referring that these kinds of practical applications are getting more pervasive and mature. These authors do also say that important existing work has been focusing on three main types of systems, concerning visualisation-based systems, co-design systems, and CE-based systems. To this end, the authors refer to collaboration as being driven by the development of logical and intelligent co ordination mechanisms to facilitate human-human and human-computer relationships.

Although, the main ideas expressed through these works also fit the very well established definition of the CE concept (Putnik and Putnik, 2019), which it is quite different from the collaboration one (Putnik and Putnik, 2019; Putnik et al., 2021b; Putnik et al., 2021c). In fact, besides the common importance of communication, alongside with the sharing issues, interaction, and interdependence or interplay, either in CE and CollE, the existence of a common goal, along with coordination, consensus, and negotiation issues are all well-known key aspects underlying CE (Putnik and Putnik, 2019) but do not have to comply with collaboration. Thus, the main existing contributions do not fully comply with the proposed CollE concept, but just with its lower or basic level, which, in fact, correspond mainly to the basic or lower level issues underlying the proposed CollE concept, and the CE one.

This is because in collaboration the existence of a common goal is not a requisite or even important, but instead, the existence of a "common understanding", in a broader sense, in order to enable and promote different points of view, and a constructive dialogue or discussion about some subject, which is enriched by diverse types of feedback and opinions that can even be opposite ones, with the main aim of reaching a common learning or co-learning stage—thus being important the existence of multidisciplinary interplaying teams, for promoting further discussion and an enriched or true co-creation and thus, innovation, which is considered a key issue in collaboration, and in the current digitalization era.

In this regard, an important contribution for the collaboration concept is put forward by Schrage, in 1990 (Schrage, 1990) in his book *Shared Minds*, in which he says that collaboration is not about agreement but about joint creation, and which thus does support the proposed CollE concept. Although, co-learning being, for us, considered the real key issue in or for enabling such co-creation, thus reaching a higher or full level of collaboration.

In (Putnik et al., 2021b,c), the existence of an important difference between the CE and collaboration, which are frequently mismatched terms or taken as synonyms, is further clarified. Therefore, there is a need for distinguishing the semantic contents of these two concepts, besides the necessity to distinguish these two from others, also more or less closely related ones, for instance, about simultaneous or parallel engineering (Putnik and Putnik, 2019). Therefore, it is of utmost importance to notice and understand that in the M&M context, collaboration is mainly driven by new, emergent, organis ational, and management concepts that refer, for instance, to new features required for the engineering design, and regarding organis ational and management issues and approaches.

According to Putnik et al. (2021b), the two new emerging theories, paradigms, and approaches that inform engineering design and practice, and on which base the definition of our CollEng is proposed, are the complexity theory and semiotics, and in particular the complexity management in organis ations and organis ational semiotics (Putnik et al., 2021b). Thus, in a broader sense, the CollE, can also be seen as a new engineering design approach (Putnik et al., 2021b; Abbass et al., 2018).

Main remarks

- Concurrent engineering can be considered a lower level of collaboration
- The human, and co-learning are fundamental to enable a higher level of collaboration
- Multidisciplinary teams enrich collaboration
- Different views, arguments and opinions, and dialogue favours or promotes collaboration

- Unlearning and dissipation can be necessary to permit true collaboration
- Small teams facilitate collaboration while easing reaching a common understanding and consensus
- Collaboration enables and promotes sustainability, based on real innovation, openness and emergence of ideas and knowledge.

Conclusion

Collaboration, is a term that has been frequently used but many times it has been mismatched with other, more or less closely related ones, for instance with concurrent engineering, which although having some similarities, is quite different in nature. This frequent misunderstanding about collaboration, namely about its application in engineering and industrial areas, has motivated this chapter, in order to contribute to the further clarification of the collaboration paradigm, based on an SLR, along with a refined proposal of a collaboration conceptual model.

The collaboration concept proposed includes two main levels, the fist one regarding the necessity to satisfy the underlying connection, communication and sharing sub concepts, and the second, higher level one, about the accomplishment of learning (co-learning), and co-creation, for enabling a full or complete level of the proposed collaboration concept, along with some kind of application or implementation, regarding its consideration in terms of collaborative engineering—manufacturing and management (Coll-M&M).

Through the study carried out, based on the SLR methodology used, it was possible to realis e the importance of the human role, not just in the context of CollE-M&M, but further in the Industry 4.0 (I4.0).

Learning was also shown to assume a fundamental importance in collaboration, namely in CollE-M&M.

It was also possible to realis e the importance of collaboration (CollE-M&M) in the I4.0, and of the digital and technological support arising from the I4.0 in promoting or enabling CollE-M&M.

References

Alavi, M., Wheeler, B. C. and Valacich, J. S. (1995). Using IT to reengineer business education: An exploratory investigation of collaborative telelearning. *MIS Quarterly*, 293–312.

Blackburn, G. S., Bilodeau, P., Cooke, T., Cui, M., Cusson, M., Hamelin, R. C. and Porth, I. (2020). An applied empirical framework for invasion science: Confronting biological invasion through collaborative research aimed at tool production. *Annals of the Entomological Society of America, 113*(4), 230–245.

Barnes, T. A., Pashby, I. R. and Gibbons, A. M. (2006). Managing collaborative R&D projects development of a practical management tool. *International Journal of Project Management, 24*(5), 395–404.

Huang, C. Y. (2002). Distributed manufacturing execution systems: A workflow perspective. *Journal of Intelligent Manufacturing, 13*(6), 485–497.

Karp, A. and Pardo, B. (2017, August). HaptEQ: A collaborative tool for visually impaired audio producers. *In: Proceedings of the 12th International Audio Mostly Conference on Augmented and Participatory Sound and Music Experiences,* pp. 1–4.

Moreno-Guerrero, A. J., Jurado de los Santos, P., Pertegal-Felices, M. L. and Soler Costa, R. (2020). Bibliometric study of scientific production on the term collaborative learning in the web of science. Sustainability, *12*(14), 5649.

Mowery, D. C. (1989). Collaborative ventures between US and foreign manufacturing firms. *Research Policy, 18*(1), 19–32.

McClellan, M. (2002). *Collaborative Manufacturing: Using Real-time Information to Support the Supply Chain.* CRC Press.

Putnik, G. D., Putnik, Z., Shah, V., Varela, L., Ferreira, L., Castro, H., Catia, A. and Pinheiro, P. (2021). Collaborative Engineering: A Review of Organisational Forms for Implementation and Operation. *In: IOP Conference Series: Materials Science and Engineering*; IOP Publishing: Bristol, UK, 2021; Volume 1174, p. 012028.

Putnik, G. D., Putnik, Z., Shah, V., Varela, L., Ferreira, L., Castro, H., Catia, A. and Pinheiro, P. (2021). Collaborative Engineering definition: Distinguishing it from Concurrent Engineering through the complexity and semiotics lenses. *In: IOP Conference Series: Materials Science and Engineering*; IOP Publishing: Bristol, UK; Volume *1174*, p. 012027.

Putnik, G. D., Škulj, G., Vrabič, R., Varela, M. L. R. and Butala, P. (2015). Simulation Study of Large Production Network Robustness in Uncertain Environment. *CIRP Ann.—Manuf. Technol., 64*, 439–442. https://doi.org/10.1016/j.cirp.2015.04.118.

Raike, A., Keune, A., Lindholm, B. and Muttilainen, J. (2013). Concept design for a collaborative digital learning tool for film post-production. *Journal of Media Practice, 14*(4), 307–329.

Sousa, A.C., Bertachini, A.F., Cunha, C., Chaves, R. and Varela, M. L. R. (2021). Literature Review and Discussion on Collaborative Decision Making Approaches in Industry 4.0, *FME Transactions, 49*(4), 817–826, https://doi.org/10.6036/9997.

Yang, W. and Takakuwa, S. (2017). Simulation-based dynamic shop floor scheduling for a flexible manufacturing system in Industry 4. *In: Winter Simulation Conference* (*WSC*). pp. 3908–3916. IEEE.

2

Relation between Collaborative Engineering and Sustainability in the Industry 4.0

A Human-centred Vision

Collaboration in the digitalization era is of utmost importance, as it enables sustainable development in Industry 4.0 (I4.0) context. Collaboration is a concept widely used, being applied in quite different contexts, varying from a more theoretical perspective to some more practical ones, and from technological scenarios up to organisational, business, manufacturing, and decision-making and management processes and practices. In this chapter a general view over different approaches and application scenarios of the collaboration term is explored, along with an overview of the main topics that have been explored worldwide correlating collaboration with sustainability in I4.0, supported by a systematic literature review.

As a result, it was possible to realise that the widened portfolio of contributions in the literature can be grouped and synthesised through a proposed framework, including three main blocks about models, tools, and business approaches about collaborative manufacturing and management in I4.0 to reach sustainability goals. It was also possible to recognize that learning, and further, co-learning are key issues correlating collaboration with the I4.0, along with the human factor, which assumes a central role for linking these two domains, which combined, are of utmost importance to enable companies to reach a sustainable development in the current digitalisation era. The main limitation of this study is related to the broad spectrum of topics underlying the three domains, thus requiring further developments, to be able to encompass detailed analysis and further correlation of them, for instance, regarding the analysis of the industrial understandings on the importance of communication, and learning, thus collaboration approaches, along with their impacts, for the promotion of a sustainable development of companies in the current digital age.

1. Introduction

Collaboration, along with sustainability concerns, regarding the social, the economic, and the environmental focuses, is not new, although assuming a new importance currently in the digital era, as has been expressed through several different authors, regarding different perspectives and aims, and some main ideas are summarised next.

The authors in (Abbass et al., 2018; Fernández-Caramés et al., 2018; Hippertt et al., 2019) mention that Human-centred design is an approach to system design and development that aims to make interactive systems more usable and useful by focusing on their use by operators and their requirements within a collaborative industrial environment. Thus, the authors argue that their proposed approach enhances effectiveness and efficiency, improves human well-being, user satisfaction, accessibility and sustainability, and counteracts possible adverse effects of use on human health, safety, and performance.

Bello et al. (2019) refers to contributions of a special section focusing on embedded and networked systems for intelligent vehicles and robots. In this work, it is mentioned that embedded and networked systems for intelligent vehicles and robots are expected to have a significant economic, societal, and technological impact on industrial and automotive applications. Further, it is said that among the aspects that will benefit from these technologies the first one is safety, thanks to the reduction of accidents caused by human errors. Moreover, another issue considered is the positive effect expected on sustainability, due to the increase in transport systems efficiency. Also, comfort and inclusiveness is mentioned to be improved, ensuring users' freedom for other activities and "mobility for all ". The authors assert that logistics and factory automation are among the main areas that will take advantage of intelligent vehicles and robots that are expected to play a key role in I4.0, where intelligent vehicles and industrial robots will move and operate autonomously and cooperatively.

To clearly expose the main contents of this chapter, it is organised as follows. In Section 2, the relation between collaboration and sustainability in Industry 4.0 will be briefly explored. Section 3 will focus on the importance of the human role in the current digital age. In Section 4, a proposed human-centred collaborative engineering management framework will be briefly presented. In Section 5, a summarised discussion on the relation between collaboration, sustainability, and I4.0 will be put forward. Finally, in Section 6, some main conclusions will be presented.

2. Relation between Collaboration and Sustainability in Industry 4.0

The term collaboration, as mentioned before, implies many different sub-terms, regarding diverse kinds of underlying collaborative processes or practices (Putnik et al., 2021a,b; Ferreira et al., 2022).

According to an extended literature review and study conducted, it was possible to reach a set of considered most relevant and closely related collaboration terms. Centred on this study, Figure 1 was created based on the set of 287 valid contributions reached, to synthesise the

Figure 1. Collaboration and some other more or less closely related terms frequently used (adapted from: Varela, Putnik, & Romero, 2023).

most relevant, best known or more frequently referred terms, besides the term 'Collaboration' itself, and other more or less similar ones, which are also used, such as: 'cooperate', 'co-work', 'co-learn', 'share', 'co-create', 'co-design', or some similar or equivalent ones, among others.

Therefore, there are some other terms, more or less closely related to collaboration, and which are also frequently used, for instance, to mention some core aspect or characteristic of a process or a practice, which is, to some extent, related to collaboration, such as: 'Communication', 'Dialogue', 'Share', 'Give', or 'Serve', among others, as shown in Figure 1.

The Importance of the Human Role

Nowadays, in I4.0, the human importance in collaboration is highlighted, as expressed in Figure 2, in which the human assumes a central position, as being considered a fundamental vehicle or element to accomplish collaboration in I4.0, besides its fundamental role, along with collaboration, as a whole, in enabling or promoting sustainability. It is considered that there is no true sustainability without collaboration (Manupati et al., 2022).

Collaboration is thus considered to be of utmost importance in the current digitalization era, along with the human role, being of prime importance in the I4.0. Besides, as mentioned in (Putnik and Putnik, 2019; Ferreira et al., 2022; Manupati et al., 2022), there is no true collaboration without learning, which also assumes a major role currently in I4.0, and the human is the fundamental vehicle to enable and promote learning, for further reaching a sustainable development of companies, as expressed through Figures 2 and 3.

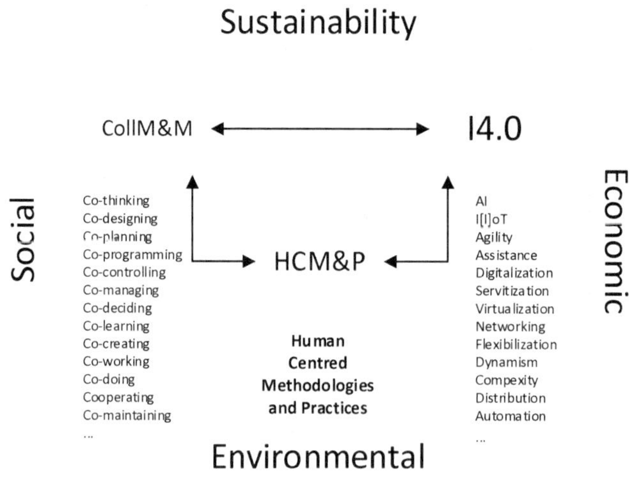

Figure 2. Correlation between CollM&M and I4.0 in the scope of sustainability (adapted from: Varela et al., 2023).

Figure 3 expresses that reaching sustainable development in companies requires collaboration, which, in turn, requires learning or co-learning, and further the human in the loop.

Additionally, in Figure 3, the human- centred practices also assume a fundamental role for enabling CollM&M and I4.0 correlations and further a sustainable development of companies, while collaborating through different kinds of collaboration processes and practices, the so-called Co-X (Sousa et al., 2021), which, for instance, include co-working, co-learning, co-designing, co-maintaining, and co-deciding, among others. All these so- called Co-X are supported and further promoted by a widened range of I4.0 approaches and technology, such as: I[I]oT, digitalization, servitization, virtualization, networking, etc.

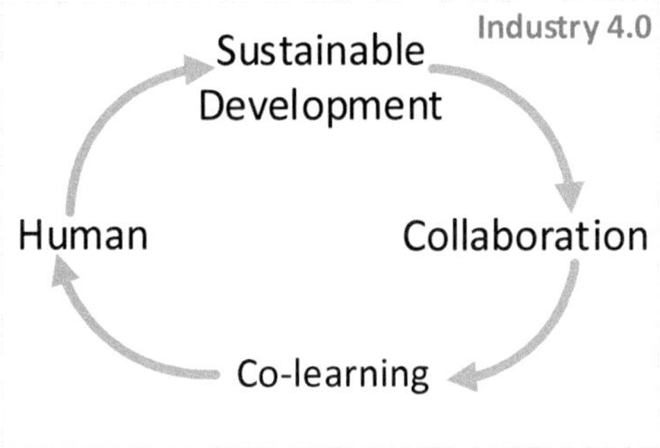

Figure 3: Relation between collaboration, learning, and the human role in I4.0 for reaching sustainable development (adapted from: Varela et al., 2023).

In this regard, it is important to highlight the central role of collaboration and the human interactions, arising through multidisciplinary scientific and technological domains, to enable and promote mutual learning and enriched theoretical and managerial developments and practices, by using suitable models, methods and approaches, through diverse kinds of means, tools, systems and platforms, to enable sustainable CollM&M in I4.0 (Liu et al., 2019a,b). Thus, a special focus relies on human-based collaboration, supported or enhanced and fostered by I4.0 technology (van Eijnatten and Putnik, 2004; Putnik and Putnik, 2019; Putnik et al., 2021a,b,c; Manupati et al., 2022).

A widened set of works do mention the importance of the human role in collaboration, particularly nowadays, in I4.0.

One frequently mentioned key aspect underlying collaboration, for instance, in Li and Qiu (2006) and Wang et al. (2002), is related to the possibility of augmenting the capabilities of individual specialists, along with the enhancement of their ability to interact with each other and with computational resources.

The authors in Li and Qiu (2006) state that a collaborative mechanism of a system is needed for a specific design along a distributed architecture to meet the functional and performance requirements imposed, through sharing diverse and complex forms of information, further supported by a multi-disciplinary design team and integrating heterogeneous application services.

The frequently mentioned parallel and synchronous characteristics, alongside the importance of interaction and multi-disciplinary issues, regarding the joint working teams, which are, in fact, fundamental ones, also in CE besides its importance in the collaboration scope.

Besides the enrichment that arises from the interaction underpinned by a multi disciplinary team in or for promoting interplay and constructive and diversified discussion between entities or stakeholders, there is no real need to force the existence of big or complex multi disciplinary teams in collaboration, because although this being important in CE, it is not a must or even important for enabling co-learning or co-creation, in CollEng. In fact, quite heterogeneous and/or complex teams can even be a problem in or for promoting collaboration, which are usually better suited or established when occurring in more contained, simpler, or in lower dimension groups of interacting people.

The authors in Longo, Nicoletti and Padovano (2017) say that I4.0 requires human operators with experience to face increased complexity of their daily tasks, requiring them to be highly flexible and to demonstrate adaptive capabilities in very dynamic and smart working environments. Therefore, there is a necessity for tools that can be easily embedded into everyday practices of operators, to enable them to combine complex methodologies with high usability requirements of the tasks.

3. Human-centred Collaborative Engineering Management Framework

Currently, in the digital era, collaborative engineering and management, centred on the human, assume a core importance for enabling a sustainable development of organisations.

Thus, another study was carried out to reach main findings regarding the state of the art about the importance that has been given to the human role, regarding collaborative methodologies and practices in the current digital era.

The reached outcomes of a deep study conducted will be summarised next, and further expressed through a general view over different approaches and application scenarios of the collaboration term, along with a set of the main topics that have been explored worldwide correlating collaboration with main I4.0 dimensions or pillars, based on an SLR.

Consequently, in this section a framework is presented for synthesising the main correlations between CollM&M and I4.0, based on three main issues, about models, tools, and business axis, due to the importance that these two scientific and technological domains assume together in the current fourth industrial revolution, further aiming at a sustainable development of companies, namely, industrial ones.

Thus, the Collaborative Manufacturing and Management Framework in Industry 4.0 is briefly described next, along with the main findings reached.

One of the most difficult problems arising in the context of engineering and industrial management science, besides the problem of data collection, is related to information processing, which, under the scope of I4.0, has to be achieved on a real-time basis. In this context, interoperable and interactive software-based systems have to support business and organisational decision-making activities, varying from a purely automated to a human interaction based decision process, relying on automatically acquired information, its processing and further analysis and use. These processes require the selection and compilation of information, according to underlying business models and objectives and within several kinds of manufacturing processes, to accurately and promptly solve engineering and industrial management problems in companies.

In this regard, it is our conviction that collaborative M&M processes and practices will enable the promotion of a successful I4.0 era, by innovating through the integration of several interdisciplinary areas, enhancing synergies between them, in order to maximise the effectiveness and efficiency resulting from those synergies.

The proposed CollM&M framework thus includes a set of three main models, regarding the development of models, the use of a set of tools, and business approaches and methodologies, as the full multidisciplinary coverage of scientific and technological domains promotes the development of methodologies, approaches, and tools for enabling a sustainable development of companies through CollM&M processes and practices in I4.0:

- models-team: from projects, models, and plans to the development of I4.0 approaches;
- tools-team: model development, specification, analysis, and validation;
- business-team: feasibility studies for the practical implementation of proposals in an industrial engineering context.

The proposed framework does thus aim at enabling deep insights and a more organised or structured analysis regarding the interrelation and mutual benefits regarding the interaction or relation between collaborative

M&M and I4.0 dimensions previously referred in Section 3, and further based on the deeper analysis carried on the subset of the considered most relevant publications in the focused domain, regarding the analyses and study performed.

In this regard, approaches oriented to the integration of collaboration concerns, along with I4.0 requirements oriented to some specific methodologies, and based on knowledge transmitted on the corresponding scientific area through fundamental methods, have to be considered or specified to develop appropriate conceptual models for specific application contexts, as illustrated through Figure 4.

The use of diverse kinds of models and tools, based on appropriate approaches, will allow putting into use collaborative M&M processes and practices, by combining different kinds of paradigms, along with appropriate research, and scientific methods to enable a strategic, multidisciplinary, and holistic vision in a focused project. It is also important to be able to further analyse and validate the specified models used, for accomplishing each particular objective regarding the application of this framework in a specific I4.0-oriented company or research context.

The models and specific tools, along with appropriate collaborative platforms, have to be conceived, oriented to some specific application domain, which will enable the analysis of underlying business models through economic analysis and feasibility studies for implementing proposals, in parallel with the application of quality analysis approaches and techniques.

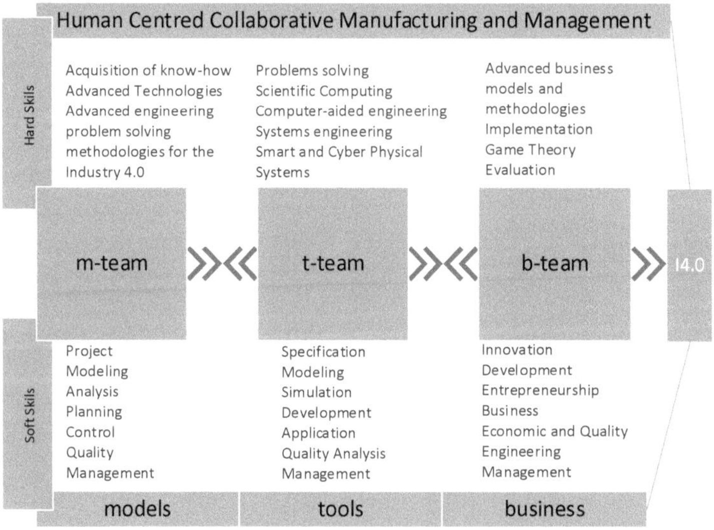

Figure 4. Human-centred collaborative framework for integrating I4.0 oriented models, tools, and business issues in the M&M domain (adapted from: Varela et al., 2023).

This collaboration framework is highly recommended to be put to work, enabling collaboration, as a central concern in the I4.0. Most proposals have a clear direction or concern in putting into practice models and tools or platforms regarding scientific and technological objectives, but other aspects, for instance related to the business itself and the underlying organisational structures, have had less visibility or expression, although having gained increased attention lately, through more recent contributions to the I4.0 concept.

Thus, although it is more or less consensual and undoubtedly that the scientific and technological domains are of utmost importance in I4.0, a higher focus and effort has been put to bring other subjects into analysis, namely, regarding contents related to business and organisational models in companies, along with the underlying tools and technologies, which are of utmost importance nowadays. Thus, within the I4.0 it is important to encompass a more significant emphasis on models covering organisational, management, and production systems aspects, in particular, while also betting on technologies and tools considered fundamental ones to fully satisfy the current requisites imposed by the I4.0, for instance, regarding the scientific area of engineering and industrial management, in which the human role assumes a fundamental relevance, promoting co-learning, and thus sustainable processes and practices in the I4.0 (Peruzzini et al., 2017; Varela et al., 2022; Putnik et al., 2021a,b,c).

In this regard, it is noticeable that fully integrated collaborative M&M processes and practices have to be further promoted, based on a deeper multidisciplinary attitude and, consequently, endowing it with an even broader and more complete nature, in order to cover, even more transversally, the whole set of main 19 I4.0 dimensions previously summarised in Section 3, and by further exploring them in the current I4.0 context, based on collaboration principles along side with I4.0 enabling technologies and tools, as intended through the proposed CollEng-I4.0 framework.

The main objective of the general collaborative framework oriented to I4.0 proposed is to put professionals in Engineering, along with the ones arising from other domains, including scientific and technological ones, and more specifically in the area of industrial and systems engineering, aware of the necessity to collaborate.

We believe that with appropriate skills and by combining them with other skills, engineers and industrial engineers will continue to play a fundamental role, not just in fulfilling specific and differentiating requirements underlying I4.0, but also in terms of integration of heterogeneous competences, for properly putting to work and manage production systems, including SP[P]S, and, in this regard, collaboration competences are fundamental to be held through multidisciplinary teams.

Therefore, the proposed collaboration framework includes three main underlying areas or axes related to models, tools, and business. This framework is a novel contribution to the field, by considering several recent domains of the I4.0, along with its integration in a collaboration perspective, based on fundamental requisites, related to the necessity of interaction, including the expected scientific and technological gains arising from the integration of other areas, such as informatics and computer sciences with areas of industrial and systems engineering, aligned with I4.0, and underlying AI requirements, among others through business strategies alignment, from design to implementation.

The proposed CollM&M-I4.0 framework further aims at raising the consciousness among professionals dedicated to the conception and development of innovative ideas and projects, from the initial idea to the placement and validation of solutions on the market. To this end, a collaborative framework for I4.0 will allow the development of multidisciplinary competencies, through the interaction of professionals from diverse areas or competencies, namely in the context of industrial companies, in order to provide a more effective and efficient transfer of knowledge and technology, promoting mutual development and valorisation, based on learning organisation principles, while contributing to the promotion of innovation, entrepreneurship, and regional, national, and international sustainable socio-economic development, while paying further full attention to environmental issues.

4. Discussion on the Relation between Collaboration, Sustainability, and I4.0

Collaborative Manufacturing and Management is a key issue in the context of a sustainable Industry 4.0, along with the emerging new paradigms and practices arising in the underlying Cyber Physical Systems.

The emergence of new technologies for communications, processing, and analysis support, together with new physical devices (sensors, controllers, actuators), increasingly able and innovative systems to explore them, and made the smart and sustainable attributes really emergent and relevant topics on manufacturing business models.

The capacity to have multiple agents working together, i.e., cooperating and, mainly, collaborating, requires the existence of dynamic and efficient interoperability practices. One of those practices is the in-time and continuous collaboration between all those expected manufacturing agents, namely humans, machines, and processes. Collaboration can be connectivity or systems integration in a technical perspective or it can be effective communication between humans-machine-humans or only humans, instead.

Furthermore, in the current I4.0, according to (Finance, 2015; Hankel and Rexroth, 2015; Xu et al., 2018), manufacturing and management (M&M) paradigms, methods, and tools should be wisely developed to fulfil sustainability issues, regarding not just economic but also social and environmental concerns by simultaneously fostering organisational efficiency excellence and transformational initiatives, therefore contributing to enduring business results (Fonseca et al., 2021).

Thus, collaboration is an important paradigm that implies an interaction between two or more entities, and though the fundamental inclusion of the human, based on multidisciplinary working teams, sustainability goals can be properly approached and favourable outcomes can be reached, either in academy or industrial and general social levels.

The term co-learn (Arrais-Castro et al., 2015; Ferreira et al., 2022; Putnik et al., 2021a,b), assumes thus a core importance to reach the sustainability requisites stated, particularly in the I4.0. In the context of collaborative M&M, this paradigm is applicable between any two or more entities, which includes not just common manufacturing resources or machines (M-M) but also humans (H-H) and their interaction with the machines (H-M) (Ferreira et al., 2022).

The I4.0 is characterised by a widened set of other paradigms, methods and tools that can promote sustainability in manufacturing systems, with flexibility as one important aspect (Reddy et al., 2017). However, this flexibility cannot just arise directly from the flexibility underlying the manufacturing resources but also from other perspectives, namely, through the flexibility subjacent to the manufacturing systems itself, while enabling quick adaptation of the production under dynamically varying conditions. These requisites arise either internally, in the factories, or from the outside, namely, regarding the need to fulfil the customers' expectations or needs (Arrais-Castro et al, 2015; Putnik et al., 2021a,b; Reddy et al., 2017; Varela et al., 2014, 2018, 2019).

Although not new, the simulation technique has also been gaining a refreshed importance nowadays in the I4.0, being one of its main pillars (Canadas et al., 2017; Putnik et al., 2015; Rodič 2017). Simulation is a very versatile technique that enables the implementation of different methods by using various software and tools, which can be easily adapted to the specific needs of the manufacturing systems (Canadas et al., 2017; Rodič 2017). One such need consists of supporting manufacturing management (Canadas et al., 2017; Rodič 2017; Varela et al., 2003; Varela et al., 2008) to reach appropriate solutions in manufacturing systems, varying from make-to-stock (MTS) to make-to-order (MTO) production philosophies, such as engineer-to-order (ETO) and design-to-order (DTO). In this regard, advanced simulation, along with other management approaches and systems, for instance, based on game theory, and other advanced and integrated optimization approaches, namely, for chaos and complexity management, based on distributed, collaborative, and real-time management principles, are of utmost importance nowadays, in the I4.0 (Alves et al., 2021; Eijnatten and Putnik, 2004).

In the context of I4.0, another important issue starts with the capability of designing smart products and advanced materials, and/or production systems based on concurrent and collaborative engineering, e.g., on open design and advanced design theory. In this regard, metatheory, formal theories, and formalisms, along with learning organisation principles, organisational semiotics, and standards, are relevant nowadays (Eijnatten and Putnik, 2004).

Key issues

- The importance of the human as the main agent to allow and promote learning (co-learning) and, therefore, as a fundamental element in the concept of collaboration, and additionally as a link that closes the circle of sustainable development of organisations in the digital age.
- There are different forms of collaboration, and require the existence of awareness, openness, flexibility, and agility to enable effective co-learning between two or more entities.
- New technologies are facilitators and support collaboration, and are therefore of paramount importance in the current digital age, further promoting and enabling a sustainable development of companies (Liu et al., 2019). However, collaborative processes and practices, in essence, do not require the existence of the new technologies underlying the Industry 4.0, in which humans are becoming increasingly more important in technology monitoring and control (Brettel et al., 2014), and it is expected a growing need for requalification of human skills, and the enrichment of its competences, towards multidisciplinarity to reach the so-called "knowledge worker" (Van Laar et al., 2017).
- Collaboration within and between organisations can occur at different levels and in different ways, ranging from design phases regarding product, production systems, processes, methods, and tools to support manufacturing and management, regarding internal and external logistics, including extended processes and interoperation among companies, and interactions with suppliers, business partners, and customers, based on appropriate business models, technology, and management paradigms and approaches through extended networks of stakeholders.
- In the considered collaboration concept, collaborating entities need to reach a common understanding for consensus building about

something (Schrage, 1990; Putnik and Putnik, 2019; Putnik et al., 2021b,c; Varela et al., 2022 a,b,c, 2023a,b).

- Collaboration is extremely important in engineering and industrial management processes and practices in the current era of digitalization, in order to guarantee and promote the sustainable development of organis ations, at the economic, social, and environmental levels, based on consciousness, and transformative change of stakeholders (Putnik et al., 2021b,c; Varela et al., 2022 a,b,c, 2023a,b).
- Several methodologies and technologies are mentioned in the literature for supporting geographically dispersed teams or entities within or between organisations that collaborate in order to share data, knowledge, tasks, and processes, namely in production (Li and Qiu, 2006; Knoben and Oerlemans, 2006; Bilberg and Malik, 2019), and an important aspect consists on maintaining small teams, as well as a contained quantity and variety of collaborative actions, as a way of facilitating collaboration processes, when trying to reach a 'consensus' since each act of collaboration has distinct requirements (Draulans et al., 2003; Putnik and Putnik, 2019; Varela et al., 2022,a,b,c; 2023a,b).
- However, there are still some concerns and difficulties to overcome in the implementation of collaborative processes and practices in organisations, namely, in terms of safety (e.g., humans co-working with machines), complexity, and security in data transferring, sharing, and processing, including appropriate procedures and tools, along with platforms for supporting the interaction in and between companies.

The importance of the human role to link collaboration processes and practices in I4.0, can be pointed out as follows:

- ➢ the human assumes a central importance to promote and enhance collaboration in engineering processes and practices, regarding collaborative manufacturing and management (CollM&M);
- ➢ sustainable manufacturing and management has to consider the three economically, social, and environmental dimensions;
- ➢ collaboration approaches can be enhanced and reinforced through the use of high scientific-technological knowledge and tools, including appropriate software and platforms for properly supporting companies and organisations in I4.0;
- ➢ transversal skills and competences have also to be further developed, through corresponding multidisciplinary teams;
- ➢ New technology can support and promote high performance collaborative engineering and sustainable processes and practices

(e.g., collaborative management) for the empowering companies, and society, in general, in the digital era.

5. Conclusion

In this chapter the main pillars of a proposed collaborative framework were presented in the light of collaborative processes, practices, tools, and platforms in the Industry 4.0, aiming to enable and promote a sustainable development of companies and society. In this regard, the main requisites of the framework for envisioning the collaborative processes and practices were explored, and shortly described and discussed, to highlight some main collaborative applications, in different kinds of contexts and domains, namely industrial ones.

The proposed framework for collaborative processes and practices in the I4.0 was thus briefly analysed, to reach a set of main issues, considered relevant, namely, in existing decision-making processes and platforms, based on an extended literature review and study conducted. Through this study it was possible to realise the value added of the proposed collaborative practice framework, to promote and reinforce collaboration and sustainability in the I4.0.

As future studies, is planned the complete development and implementation of the proposed framework, for being further applied in real life manufacturing application scenarios.

Summarising, this study did enable us to clarify the importance of collaboration, for instance CollEng-M&M, in the current digitalization era, in order to promote a sustainable development of companies, namely industrial ones, regarding not just economical, but also social and environmental issues. In the current I4.0-oriented CollEng-M&M context the essential importance of the human in the loop was clarified, along with its learning capabilities, for which the ICT-oriented support, subjacent to the I4.0, was also visible for enabling and promoting collaboration.

Besides the awareness of the importance of collaboration (CollEng-M&M), some difficulties or concerns do persist nowadays, for its successful practical implementation in companies, namely in industrial ones, for instance regarding human-machine or more specifically human-robot collaboration, mainly to ensure appropriate and secure working conditions, among other concerns and restrictions, which have to be further explored and focused, as this kind of collaboration has been gaining a primer importance in the current digitalization era, and has been already mentioned to be also a further main pillar in the Society 5.0.

Based on the study conducted, it was thus possible to reach a set of clusters of key terms that did enable us to synthesise the proposed CollM&M framework, which integrates three main blocks about models, tools and business approaches.

The proposed framework enables to integrate and synthesise the main issues regarding collaborative M&M processes and practices, in the I4.0. This is an important contribution, due to the relevance that these two scientific and technological domains assume nowadays, when combined, to further enable a sustainable development of companies, namely in the industrial context.

As previously exposed, the concurrent and collaborative engineering concepts are frequently mismatched in the literature, namely by imposing the existence of a common goal, for instance, between two or more companies, while cooperating in some kind of manufacturing or management process. Although, as mentioned in (Putnik and Putnik, 2019; Putnik et al., 2021a,b,c; Ferreira et al., 2022; Manupati et al., 2022; Varela et al., 2022), and further described in the first chapter, this is actually one main difference that enables to distinguish both concepts or M&M paradigms, and further promoting innovation, by the emergence of new approaches and solutions, thus contributing to the enrichment of interactions or inter play among companies, and thus, promoting their sustainable development, based on the central human role, and its learning/co-learning potential, which can further be enriched through the fundamental support provided by the current approaches and technology underlying the I4.0.

The main limitation of this study it related to the broad spectrum underlying the combination of subjects arising from both domains (CollM&M and I4.0) combination, which was studied based on literature available through the b-on database, thus requiring further developments, to enable covering a broad range of applications scenarios and its detailed analysis, namely through some case studies, to reach additional conclusions and a full research concept validation. Such a deed analysis will further permit analysis of industrial understandings about the importance of collaboration issues, along with the importance underlying communication processes and supporting Industry 4.0 technology, among other issues.

References

Alves, C., Putnik, G. D. and Varela, M. L. R. How Environment Dynamics Affects Production Scheduling: Requirements for Development of CPPS Models. *FME Trans.*, 2021, *49*, 827–834. https://doi.org/10.6036/9997.

Arrais-Castro, A., Varela, M. L. R., Putnik, G., Ribeiro, R. and Dargam, F. Collaborative negotiation platform using a dynamic multi-criteria decision model. *Int. J. Decis. Support Syst. Technol.*, 2015, *7*, 1–14. https://doi.org/10.4018/ijdsst.2015010101.

Canadas, N., Machado, J., Soares, F., Barros, C. and Varela, L. Simulation of cyber physical systems behaviour using timed plant models. Mechatronics, 2018, *54*, 175–185. https://doi.org/10.1016/j.mechatronics.2017.10.009.

Eijnatten, F. V. and Putnik, G. D. Chaos, complexity, learning, and the learning organis ation: Towards a chaordic enterprise. *Learn. Organ.*, 2004, *11*, 418–429.

Ferreira, L., Punik, G., Varela, L., Manupati, V. K., Lopes, N., Cunha, M., Alves, C. and Castro, H. A Framework for Collaborative Practices Platforms for Humans and Machines in Industry 4.0 oriented Smart and Sustainable Manufacturing Environments. *In: Smart and Sustainable Manufacturing Systems for Industry 4.0*; Manupati, V. K., Goran, D. P. and Rocha, M.L. (Eds.). CRC Press, Taylor & Francis Group: Boca Raton, FL, USA, 2022; in press.

Finance, A. T. C. C. Industry 4.0 Challenges and solutions for the digital transformation and use of exponential technologies. *Finance. Audit. Tax Consult. Corp.* Zur. Swiss, 2015, 1–12.

Fonseca, L., Amaral, A., and Oliveira, J. Quality 4.0: The EFQM 2020 Model and Industry 4.0 Relationships and Implications. *Sustainability*, 2021, *13*, 3107. https://doi.org/10.3390/su13063107.

Hankel, M. and Rexroth, B. The reference architectural model industrie 4.0 (rami 4.0). *ZVEI*, 2015, *2*, 4–9.

Putnik, G. D., Putnik, Z., Shah, V., Varela, L., Ferreira, L., Castro, H., Catia, A. and Pinheiro, P. Collaborative Engineering: A Review of Organisational Forms for Implementation and Operation. *In: IOP Conference Series: Materials Science and Engineering*; IOP Publishing: Bristol, UK, 2021; Volume 1174, p. 012028.

Putnik, G. D., Putnik, Z., Shah, V., Varela, L., Ferreira, L., Castro, H., Catia, A. and Pinheiro, P. Collaborative Engineering definition: Distinguishing it from Concurrent Engineering through the complexity and semiotics lenses. *In: IOP Conference Series: Materials Science and Engineering*; IOP Publishing: Bristol, UK, 2021; Volume 1174, p. 012027.

Putnik, G. D., Škulj, G., Vrabič, R., Varela, M. L. R. and Butala, P. Simulation Study of Large Production Network Robustness in Uncertain Environment. *CIRP Ann.—Manuf. Technol.*, 2015, *64*, 439–442. https://doi.org/10.1016/j.cirp.2015.04.118.

Reddy, M. S., Ratnam, C., Agrawal, R., Varela, M. L. R., Sharma, I. and Manupati, V. K. Investigation of reconfiguration effect on makespan with social network method for flexible job shop scheduling problem. Comput. *Ind. Eng.*, 2017, *110*, 231–241. https://doi.org/10.1016/j.cie.2017.06.014.

Rodič, B. Industry 4.0 and the new simulation modelling paradigm. *Organizacija*, 2017, *50*, 193–207.

Varela, M. L. R., Aparício, J. N. and Silva, S. C. A Web-based application for manufacturing scheduling. *In: IASTED International Conference on Intelligent Systems and Control*; Salzburg, Austria, 2003; pp. 400–405, 25th to 27th of June 2003, ISBN: 0889863555.

Varela, M. L. R. and do Carmo Silva, S. An ontology for a model of manufacturing scheduling problems to be solved on the web. *In: International Conference on Information Technology for Balanced Automation Systems*; Springer: Boston, MA, USA, 2008; pp. 197–204.

Varela, M. L. R., Putnik, G. D., Manupati, V. K., Rajyalakshmi, G., Trojanowska, J. and Machado, J. Collaborative manufacturing based on cloud, and on other I4.0 oriented principles and technologies: A systematic literature review and reflections. *Manag. Prod. Eng. Rev.*, 2018, *9*, 90–99. https://doi.org/10.24425/119538.

Varela, M. L. R., Putnik, G. D., Manupati, V. K., Rajyalakshmi, G., Trojanowska, J. and Machado, J. Integrated Process Planning and Scheduling in Networked Manufacturing Systems for I4.0: A review and framework proposal. *Wirel. Netw.*, 2019, *27*, 1587–1599.

Varela, M. L. R. and Ribeiro, R. A. Distributed manufacturing scheduling based on a dynamic multi-criteria decision model. *In: Recent Developments and New Directions in Soft Computing*; Springer: Cham, Switzerland, 2014; pp. 81–93.

Xu, L. D., Xu, E. L. and Li, L. Industry 4.0: State of the art and future trends. *Int. J. Prod. Res.*, 2018, *56*, 2941–2962.

3

Contextualization of Collaborative Engineering

Manufacturing and Management in the Industry 4.0

Collaborative engineering–manufacturing and management—is crucial in the current digitalization era, to permit and promote a sustainable development of companies, either traditional or extended and networked organizations, including cyber physical systems.

This chapter aims at putting forward higher-quality technical-scientific content about fundamental methodologies, models, methods, tools, and platforms about collaborative engineering, to support manufacturing and management processes and practices, aligned with the current requirements underlying the Industry 4.0, and the Society 5.0 principles and aims.

To this end, this chapter focuses on the exploration and application of collaborative management paradigms about dynamic, distributed, integrated, intelligent, predictive, parallel, and real-time based approaches and tools to enable collaborating entities, including suppliers, business partners, and other stakeholders, to develop projects and solve problems that are becoming increasingly more complex and challenging currently. Such collaborative processes and practices require companies and underlying stakeholders to be connected, and to further communicate, and share data, problems, and expertise, and other kind of resources, along with concerns, difficulties, and challenges, requiring co-learning, and the co-creation of knowledge, processes, methods, and systems to interactively support projects and problems solving.

In this chapter, advanced ideas and works compendium supported by literature review studies, along with dissertation and Ph.D. work, besides other case studies will be summarily provided, to assist scientists, practitioners, and students in high standard manufacturing management processes and practices, to properly handle their daily base problems and challenges, with a special focus on the use of recent paradigms and tools to support manufacturing management decision- making, through innovative methodologies and approaches for permitting researchers to learn, develop further work, and become advanced practitioners and promoters of collaborative management.

1. Introduction

Industry 4.0 represents the current trend in the manufacturing industry, where manufacturing, and management is to be performed within a highly connected and smart environment and in a collaborative manner, for enabling improved and sustainable companies, and society.

Diverse literature studies were carried out by various researchers, including master and Ph.D. students, to realize the importance of collaboration in the current digitalization era, on one hand, and on the other about the importance of recent approaches and technology for enabling or promoting collaboration. Main current practices of human-centered and more autonomous machine-machine approaches and applications of collaboration in engineering, for instance, in manufacturing and management domains, which will be shortly described in this chapter, along with main difficulties and further open research opportunities on collaboration.

To properly contextualize collaborative engineering, manufacturing, and management (CollE-M&M) in the Industry 4.0, a systematic literature review (SLR) was performed with the additional purpose to further correlate CollE-M&M methodologies and practices with I4.0 principles, approaches, and technology, including models, methods, and tools, based on a set of considered most relevant publications found in the literature, which were selected according to its relevance in the context of the underlying collaboration concept defined. Thus, the main research question focused in this study was the following: "What are the main relations between CollE-M&M and I4.0?"

To arrive at an answer or, at least, to bring some insights into consideration about potential correlations between CollE-M&M and the I4.0, this chapter will be organized as follows:

In Section 2, some main considerations about collaborative manufacturing and management in the I4.0 will be explored. In Section 3, a synthesis of the main I4.0 topics approached on collaborative manufacturing and management will be presented. Further, Section 4 summarizes the main collaboration dimensions in the I4.0. In Section 5, a final discussion on collaborative manufacturing and management in the I4.0 is provided, and finally, in Section 6, main conclusions are provided along with some insights regarding open future research on the focused domain.

2. Collaborative Manufacturing and Management in the Industry 4.0

To explore the main contents underlying this chapter, some information about main approaches, models, methods, and tools, including traditional systems, alongside platforms for enabling collaborative engineering, processes, and practices in the current I4.0 will be synthesized. This synthesis and analysis is considered important nowadays, to enable a proper understanding about the correlation of the collaboration and Industry 4.0 domains, which are both of utmost importance in the current digitalization era, namely, in the industrial (M&M) context, and for promoting a sustainable development of companies.

In Bechtold and Lauenstein (2014), it is referred to the current industrial revolution as being characterized by the cooperation or collaboration of intelligent machines, storage systems, production systems, and people into intelligent networks, merging the real and virtual worlds through cyber-physical systems (CPS). The authors further state that these CPS integrate information technology (IT) systems with mechanical and electronic components connected to online networks that allow the communication between machines in a similar way to social networks, and these innovative technologies enable factories to become 'smart' resulting in productions of customized products on an industrial scale while providing many opportunities for improvements in operational flexibility and efficiency, including with human work, as further mentioned in Kaasinen et al. (2020); Varela et al. (2021); and Putnik et al. (2021a).

In Louw et al. (2018), it is referred that the FoF will make use of actuators, sensors, and CPS to provide an environment in which humans, machines, and resources will communicate as in a social network. The authors consider information flow a key enabler of such FoF. Further they state that industrial engineers, as designers and improvement agents of such FoF, will need to develop better skills in various aspects of data analytics and information communication technologies.

Moreover, the authors in Bechtold and Lauenstein (2014) mention that CPS' influence on the human factor is linked through four elements: (1) tools and technologies, (2) organization and structure, (3) working environment, and (4) organizational cooperation. They consider that the FoF will increase the need for skilled digital work, that there will be a decrease in the need for manual work, and that the workers will be provided with the exact information needed, in real time (RT), to perform properly and efficiently execute tasks. Thus, intelligent systems will further make it possible for the worker to make qualified decisions in a shorter time. Also that CR will share a workstation with humans, and that these robots will support them, for example, in situations that are critical regarding ergonomic conditions. Besides that, intelligent tools and technologies will become more autonomous and automated, but the supervision and efficient application of machines by humans will become more important than ever before.

As referred in Chen et al. (2016), intelligent industrial ecosystems enable the collection of massive data from various devices (e.g., sensor-embedded wireless devices) dynamically collaborating with humans (Varela et al., 2018). According to the authors this is essential to improve the efficiency of industrial production/service. In this paper, the authors propose a collaborative sensing intelligence framework, combining collaborative intelligence and industrial sensing intelligence, which they state facilitates the cooperativity of analytics by integrating massive spatio-temporal data from different sources and time points.

In Petrillo et al. (2018), the authors believe and state that a significant change in the used technologies should and will proceed jointly with a significant change in organization and structure of companies. In this regard, the authors mention that workers will be capable of working in accordance with dynamically available and updated information through more or less complex data flows that will no longer be necessarily bound or restricted to a certain production area.

Thus, the new operator skills will be of prime importance to improve job management by making it more qualified, responsive, and a better informed DM process, taken remotely. According to the authors, the future working environment will be an open and creative space. Work will be more flexible and transparent, more planned, and balanced. The authors believe that the homework will increase. However, modern assistant systems will provide the workers with the ability for quick DM despite the increased complexity of their job contents. Moreover, the authors do further state that the work will be improved with respect to ergonomics. More precisely, those non-ergonomic processes are likely to become automated, to improve the workers' conditions. In the FoF, intra-organizational cooperation and communication will be fundamental. Networking and interconnectedness are focal components of the I4.0. Workers will collaborate and communicate in real time without borders using smart devices. The Internet provides the possibilities to meet globally in virtual rooms at almost any time and to reach out for required information as needed. All kinds of information and data will be ubiquitous and at the fingertips of the workers leading to a whole new level of knowledge management. Humans will communicate with each other and with intelligent machines, and intelligent machines will also communicate with each other.

Different kinds of CollEng-M&M approaches and platforms have been put forward during the last decade, and with a refreshed and reinforced importance nowadays, in the I4.0, for enabling human-centered collaboration. In this context, Computer Supported Cooperative or collaborative work (CSCW) is gaining new importance and expression (Putnik et al., 2020; Putnik et al., 2021d).

In Li and Qiu (2006), it is stated that collaboration is to establish an effective communication channel between the upstream design and downstream manufacturing to enrich the principles and methodologies to link diversified engineering tools dynamically. Thus, according to the authors, the future trends for the collaborative systems include, although not being limited to, the integration of various collaborative manners and systems.

Their proposed integral system can support interrelated activities and share domain knowledge between designers and systems to improve design quality and efficiency. It integrates modules for hierarchical collaboration that can be wrapped as services for remote revoking. The authors state that their system enables scheduling and co ordination, which they consider becoming more crucial and challenging, and to be enhanced through the use of distributed intelligent algorithms and technologies such as MASs or Web services for increasing the potential of collaboration.

The authors further mention that research and development have been actively carried out to develop technologies and methodologies to support collaborative design (CD) and development systems, and that software vendors have quickly realized the huge business opportunities in this area, having been launching to the market a variety of commercial systems to promote collaboration (Li and Qiu, 2006).

Therefore, in the scope of collaborative M&M, many different group decision-making processes can be considered, based on different kinds of models or methods and tools, which can further be put forward through different kinds of platforms or systems, frequently referred as collaborative platforms, networks, or systems (Vafaei et al., 2019).

The so-called collaborative platforms (CP) have been evolving fast, during the last decade, and can be further improved in the Industry 4.0 (or I4.0) context (Hankel and Rexroth, 2015; Arrais-Castro et al., 2018; Ferreira et al., 2022). This is due, on one hand, to the rapid development and spread of the so-called exponential technologies, along with high performance computational capabilities (Ferreira et al., 2022).

In the manufacturing environments, the Cyber Physical [Production] Systems (CP[P]S) nowadays play a fundamental role in the context of I4.0 (Hankel and Rexroth, 2015), and provide a basis for promoting CollE, through [I]IoT ([Industrial] Internet of Things), along with other underlying technologies, such as RFID (Radio Frequency IDentifiers), the Cloud, Augmented, Virtual and/or Mixed Reality, and digital twins, along with smart objects, and many different kinds of sensors, devices, and tools (Lin et al., 2015; Arrais-Castro et al., 2018; Ma et al., 2019). These tools include software for providing and processing varying types and amounts of data, namely big data, in a real-time basis, in order to fully enable integration of data and processes, and enhance the interoperability among functions and systems, which is of utmost importance for establishing collaborative practices (Manyika et al., 2011; Arrais-Castro et al., 2018; Varela et al., 2018; Ferreira et al., 2022).

Regarding the use of approaches for carrying out joint decision-making processes, different kinds of approaches can also be considered, for instance, based on group decision-making. For this purpose, several methods and tools exist for enabling a collaborative decision making process, including different kinds of multi-criteria methods, different data normalization techniques, and diverse models and tools for data processing and analysis, namely, from the AI domain, such as fuzzy data processing (Arrais-Castro et al., 2018; Vafaei et al., 2019).

The AI- based approaches, models, and tools for collaborative decision-making frequently fall within the context of the general data science domain, and are often based on machine or deep learning algorithms, as well as on fuzzy decision-making approaches, among others (Sousa et al., 2021).

According to Thomas and Kellogg (1989), human-machine symbiosis in the AI era, and especially in I4.0 environments, is at its early stages and there are still many unexplored opportunities. I4.0 enables new types of interactions between operators and machines (Emmanouilidis et al., 2019; Rauch et al., 2020). This allows a paradigm shift from independent automated and human decisions towards a human-AI symbiosis, characterized by the collaboration of AI and human intelligence (Romero et al., 2015, 2016, 2017; Guerin et al., 2019).

The authors in Bousdekis et al. (2020) put forward a Human Cyber Physical System (HCPS) framework for Operator 4.0 –AI Symbiosis and its main architectural building blocks. Operator 4.0 is defined as being an "operator of the future", a smart and skilled operator who performs "work aided" by machines if and as needed in an I4.0-oriented environment (Romero et al., 2015, 2016, 2017). The HCPS concept, building on top of human automation interaction (Hancock et al., 2013), aims at studying the symbiosis between humans and AI, in which the human is an integral part of the CPS.

Further, in Zolotová et al. (2020) is also presented an HCPS, which the authors consider important for fulfiling the current new demands for productivity and effectiveness in production. The authors state that a traditional operator is being transformed to the Operator 4.0, and in their paper, they describe evolving roles of the operators in the factories, by mentioning different ways to enhance the operators' physical, sensing, and cognitive capabilities, that according to them can be used individually or in combination to put humans into the center of the current technological revolution.

There are many other different kinds of technology and approaches that can be further explored for carrying out some collaborative process or practice, such as those relying on game theory, learning factory, and social network or community-based approaches, tools, and platforms (Reddy et al., 2017; Schuhmacher et al., 2017; Vafaei et al., 2019).

Based on SLR carried out (IJMSEM 2023), a subset of the 82 most relevant publications identified in the 5th and last stage of the SLR methodology applied enabled us to further organize them according to the main focus regarding models, methods, tools, and business strategies, approaches, and underlying technology, regarding the relation between CollM&M and I4.0, according to the seven main clusters obtained and previously illustrated through Figure 4, and the information is synthesize in Table 1, where each contribution identified in the literature is identified by its title and year.

In the I4.0 context, according to the study carried out, namely, through the deeper analysis performed on the set of 82 publications considered the most relevant ones, synthetized in Table 1, it is possible to realize that there is currently an ample space and opportunity for conducting collaborative M&M processes and practices that maximize advantages provided through the use of models, tools and general technologies, providing additional capabilities and knowledge, along with best suited business and organizational models, to be considered on each particular company and industrial application context.

Typically, regarding the specific M&M scope oriented to I4.0, some main pillars or dimensions were emphasized, such as the CP[P]S, the i[I]oT, horizontal and vertical integration, cybersecurity, additive

Table 1. Synthesis of the five main clusters of publication/terms correlating CollM&M and I4.0 to enable a sustainable development of companies (through models, approaches, tools, business strategies, and underlying technology (adapted from: Varela et al., 2023).

Main clusters of key terms
1) (in dark green): Machine/Deep Learning, neural networks, advanced optimization, models, and systems
2) (in dark blue): Big data analytics, Data science, IoT, Blockchain, and other architectures, networks, cloud, and grid computing, integrated systems, and other data-driven approaches and technology for supporting knowledge management, and human-centered activities.
3) (in red): Integrated and concurrent design, trust, performance and impact evaluation, frameworks, and technology, Simulation, digital twin, augmented, virtual and mixed reality, game theory, chaos and complexity management, and other decision-making support approaches, and tools.
4) (in light green): Artificial Intelligence, including multi-agents, cognitive, and patterns recognition approaches, along with meta heuristics, fuzzy logic, etc., HPC, computer vision, automation, CP[P]S, smart factories, collaborative robots, other I4.0-based approaches and technology, for enabling high performance manufacturing and computing, based on flexible electronics, and 2D materials, among others.
5) (in purple): Sustainability, and companies' life cycle assessment, Business models and strategies, Integration, flexibility, communication networks, frameworks, and other technology

manufacturing, advanced robotics, exponential technology, Artificial Intelligence (AI), big data, data analytics (data science), advanced simulation, along with high performance computing, and other specific scientific areas directly related to industrial engineering, such as open design, open innovation, metatheory, formal theories and formalisms, learning organization, organizational semiotics and norms, new business models, circular economy, finance and risk assessment and management, organizational change and transformation, employees, competencies and culture management, along with Cloud computing, M&M, as well as advanced interfaces, virtual and augmented reality, digital twin, business intelligence, advanced, integrated and intelligent supply network management, project and business management, interoperable and integrated manufacturing, management, quality, and maintenance systems, advanced ERPs and MES, game theory, and other advanced and integrated optimization systems, e.g., for chaos and complexity management, and also advanced energy collection and storage and decarbonization models and policies, which are relevant subjects in the context of I4.0 oriented collaborative M&M.

3. Synthesis of the Main Industry 4.0 Topics Approached on Collaborative Manufacturing and Management

Collaboration M&M issues have been widely addressed in the context of I4.0, namely, through AI-based approaches, such as those based on machine and deep learning, neural networks, big data, and other data analytics.

Moreover, many other technologies, approaches, methods, and tools have been explored, for instance, based on simulation, game theory, digital twins, and many other communication and information technologies with more or less intelligence capabilities, as illustrated in Figure 3. Figures 3 and 4 were created by using the VOS viewer software, based on the literature search process carried out based on the SLR methodology previously mentioned.

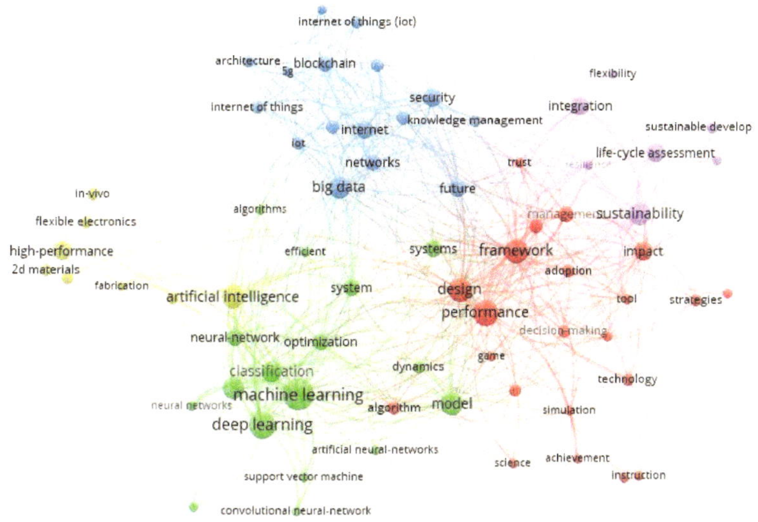

Figure 1. Main key terms correlating Collaboration and I4.0 (adapted from: Varela et al., 2023).

Source: Main key terms correlating Collaboration and I4.0 (adapted from: Varela et al., 2023).

The VOS viewer highlights the graphical representation of bibliometric maps, and, therefore, is particularly suitable to illustrate large bibliometric maps in an easy-to-interpret way (Arici et al., 2021), and can be used to construct maps of keywords based on co-occurrence data (Eck and Waltman, 2007).

In Figures 3 and 4, the size of the nodes has to do with the frequency with which the respective parameters appeared. The links between nodes represent the simultaneity of the presence of the respective terms. Thus, in Figure 4, it represents the simultaneity of the presence of terms in the same document. In Figure 4 can be seen seven main clusters of publications, grouped by the underlying focused domain, expressed through different colors.

The most impacting cluster is the one represented in dark green, mainly focused on machine/deep learning models and systems. The next impacting one is represented in purple, about the design and performance evaluation of life-cycle assessment strategies, environments, business models, and ecosystem services for reaching sustainable development in companies. Next appears the cluster expressed in red, which represents decision and communication support technology, including platforms, frameworks, and simulation models and tools. The following impacting cluster, in dark blue, refers to advanced architectures, methodologies, and models based on diverse kinds of recent approaches, namely, based on game theory and cloud and complexity management. Another cluster, marked in light green, is related to other AI-based approaches, based on neural networks and pattern recognition, along with high-performance computing and computer vision technology, among others. This cluster is focused on data- science- centered contributions, and further based on IoT, besides integration of hard and software tools for enabling flexible and interactive decision support, for instance, through digital twins, and big data analytics, besides other algorithms, for patterns and performance analysis.

Industry 4.0, Industrie 4.0, or I4.0 for short (Kagermann et al., 2013; Hankel and Rexroth, 2015) is nowadays a hot topic that is getting increased attention worldwide, namely, in the manufacturing and management scope (Hopkins and Siekelova, 2021; Ferreira et al., 2022). The industrial world revolves around production systems and is currently witnessing this new experience, the 4th industrial revolution (I4.0).

This new industrial revolution has gained an accelerated pace, and it includes the emergence of new business models (Ibarra et al., 2018).

This could be further detailed by referring to the contribution of novel Business Models, such as the EFQM 2020, that add a strategic and technologically unbiased perspective to Industry 4.0 and digital transformation. These business models provide an integrated framework aiming simultaneously to deliver performance and ensure transformation, creating enduring value for its key stakeholders and achieving remarkable results (Fonseca et al., 2021; Fonseca, 2022; Murthy et al., 2021).

New technologies play a crucial role in I4.0, namely, the so-called exponential technologies, which are gaining visibility along with new concepts and principles regarding new models, methods, means, tools, and platforms for performing production management through improved and more effective and efficient manufacturing systems configurations and models, e.g., CP[P]S (Lee et al., 2015; Rodič, 2017).

CP[P]S are being supported by and enable support to new business and management models through digital networks for a more accurate and real-time based data, information and knowledge exchange and processing (Hopkins and Siekelova, 2021; Ferreira et al., 2022), by connecting manufacturing systems with suppliers, customers, and extended sets of stakeholders, and business partners worldwide through networks, virtual or extended enterprises, companies or organizations (Vafaei et al., 2019). Improved communication means automated data acquisition, processing, and presentation/ transferring devices, tools, and systems also play a fundamental role for enabling collaboration (Vafaei et al., 2019; Ferreira et al., 2022).

Moreover, regarding the relationship between smart factory performance, IoT sensing networks, and CPS-based manufacturing in Industry 4.0-based collaborative engineering interesting recent contributions are put forward (Nica and Stehel, 2021; Novak et al., 2021; Zvarikova et al., 2021) for implementing and enabling support to new business models and management tools.

Besides, other main issues related to M&M in I4.0, for instance, occurring in the AI domain, regarding the relationship between deep learning-assisted smart process planning, digitized mass production, and smart manufacturing big data in I4.0-based collaborative engineering is further focused in Kovacova and Lăzăroiu (2021); Riley et al. (2021); and Hopkins and Siekelova (2021).

An overview of Industry 4.0

The I4.0 concept is based on digital transformation in traditional production and management methods with the introduction of information technology and it is, according to the identification of Deloitte (2014a), composed of four fundamental characteristics: vertical integration, horizontal integration, end-to-end engineering, and orchestration of the value chain by people, which assumes a central role and importance (Hankel and Rexroth, 2015; Deloitte, 2014b).

The systems that exist in a factory environment can be integrated at five levels. The integration of operational data with business data can be aligned using the ANSI/ISA-95 ISO/IEC-62246 standard "Enterprise-Control System Integration" (ISA-95) (Prades et al., 2013). This standard establishes the terminological and functional basis, good practices, workflows, data flows, and alignment between business systems, e.g., ERP, and operational control systems, e.g., MES, SCADA (and IoT and CPS middleware), including IIoT communication protocols, among others (Prades et al., 2013; Lin et al., 2015; Soldatos et al., 2016; Li et al., 2017).

Global value chain networks are optimized networks that provide real-time information about geographically dispersed factories facilitating global management and optimization through extended and globally distributed resource markets (Arrais-Castro et al., 2018; Vafaei et al., 2019). This exchange of information and resources increases transparency between factories and business partners, and promotes a high level of integration, interoperability, flexibility, distributivity, virtuality, and agility to respond quickly to varying kinds of requests about specific issues, problems or failures (Arrais-Castro et al., 2018; Varela et al., 2018a,b).

The shared information ranges from inbound logistics to storage, production, marketing, sales, and outbound logistics. In this sense, the history of each product or raw material is recorded and can be accessed through the factory system and the state of the situation can be shared with other factories, ensuring constant traceability (a concept known as "product memory") (Carvalho et al., 2016).

It is in the layer of actuators and sensors that a large part of the factory information, namely from the factory floor, is located. This very low-level information is then used by other systems (as suggested in ANSI/ISA-95) (Chen 2005; Prades et al., 2013).

In this sense, the use of protocols adopted worldwide such as MQTT, CoAP (Constrained Application Protocol), AMQP (Advanced Message Queuing Protocol), HTTP/2 (Updated version of Hypertext Transfer Protocol), IPv6 (Internet Protocol Version 6) or 6LoWPAN (IPv6 over Low power Wireless Personal Area Networks) is an accepted and appropriate practice in the implementation of the I4.0 paradigm (Kagermann et al., 2013; Hankel and Rexroth, 2015; Smit et al., 2016; Xu et al., 2018).

Despite being a relatively recent concept, efforts to standardize it have already been made in the context of I4.0, which allowed the emergence of a reference architecture. This architecture was defined by the Industrial Internet Consortium (IIC), and it is called the Industrial Internet Reference Architecture (IIC, 2017) (Lin et al., 2015; Li et al., 2017; Liao et al., 2018). Present in this architecture are concepts related to an Industrial Internet environment and its interconnections from four perspectives: business, use, functional and implementation (Lin et al., 2015; Li et al., 2017).

The Industrial Internet Reference Architecture (IIRA) integrates a security policy for manufacturing infrastructures, hardware, software, and communication, across the four perspectives presented in Lin et al. (2015) and Li et al. (2017). Another equivalent initiative is called the Reference Architecture Model Industry 4.0 (RAMI4.0), referred to in Hankel and Rexroth (2015). This architecture defines hierarchies for the development of a unified model of all components of the I4.0 present in the value chain. These hierarchies refer to the business, functional, information, communication, integration, and asset layers (Hankel and Rexroth, 2015).

The C[P]PS and smart factories, based on intelligent sensing systems, open systems, and networked and distributed manufacturing systems, as well as urban production systems, along with virtual organizations and open systems, also play fundamental roles nowadays (Canadas et al., 2017; Alves et al., 2021; Lee et al., 2015; Putnik et al., 2021; Shah and Putnik, 2019). In such advanced manufacturing systems, integration, distributivity, virtuality, agility, servitization, digitalization, and decentralization are also major issues in sustainable and collaborative processes and practices in the I4.0. In this regard, the [I]IoT, smart and ubiquitous networks based on the cloud, enable large and complex networks and their digitalization (Varela and Ribeiro, 2014; Varela et al., 2019; Li et al., 2017; Liao et al., 2018). Decisions and related actions must be taken quickly and are supported by accuracy monitoring systems (Costa et al., 2021).

Cloud-based computing, manufacturing, and management are thus fundamental currently for fully proving enhanced flexibility and suitability for enabling collaboration and effective engineering practice (Ferreira et al., 2022; Varela et al., 2018). Cumulatively, horizontal and vertical integration among partners, factories, suppliers, customers, and other businesses and/or stakeholders is also crucial in the current I4.0 era (Arrais-Castro et al., 2015).

Additive manufacturing or 3D printing also consists of other critical enabling technology and principles for promoting collaborative processes and practices between stakeholders in networked manufacturing environments. Moreover, exponential technology and advanced processes, high-performance computing, and disruptive technologies (e.g., automation and robotics, autonomous and collaborative robots, advanced mechatronics, micro and nano manufacturing, and supercomputing) are also key enablers for sustainable manufacturing and management in the current I4.0 context.

Advanced interfaces, virtual and augmented reality, and digital twin, promoting and enhancing collaboration between entities, are also critical today. These technologies enable advanced and integrated decision support systems (DSS) and databases (DB), knowledge engineering and knowledge bases (KB), automatic data acquisition, and a semantic web for enhancing collaboration.

There are also other relevant approaches, methods, and techniques in the I4.0, for instance, based on AI, e.g., machine learning and deep learning, pattern recognition, blockchain, and other technologies and methodologies for enabling and enhancing manufacturing and management (Putnik et al., 2021; Shah and Putnik, 2019). In addition, business intelligence, big data, and data analytics in the specific data science domain are essential pillars of I4.0 (Manyika et al., 2011).

Moreover, it is also of utmost importance to explore new business and organizational models, attending to the need inherent to the circular economy, finance, and risk management in and between organizations (Prades et al., 2013), that can support organizations simultaneously managing the present and transforming for the future while responding to the challenges and opportunities of changing business environments (Fonseca, 2000). The organizational change and transformation are now mandatory, regarding employees, their competencies, and culture, to reach suitable manufacturing and management (Prades et al., 2013). For industry, the management of I4.0 is a crucial issue and should also be carried out considering factors of production, directly or indirectly, in order to improve their performance (Putnik and Ávila, 2021).

All these issues are crucial for enabling advanced, integrated, and intelligent supply networks, projects, businesses, and their integrated and fully supported management, to reach manufacturing and management while ensuring high-quality standards, extended to other practices, e.g., maintenance and control (Fonseca et al., 2020). To this end, several different kinds of performance measures and goals should be considered to reach sustainability, organizational and machine robustness, scalability systems, and other advanced enterprise information systems (EIS), such as enterprise resource planning (ERP), manufacturing execution systems (MES), and systems for supply chain management (SCM).

4. Main Collaboration Dimensions in the Industry 4.0

A varying set of the main I4.0 dimensions, pillars and/ or scientific areas or domains, and underlying technology, have been put forward during the last years, as mentioned, for instance, in Kagermann et al. (2013); Deloitte (2014a,b); Hankel and Rexroth (2015); Ibarra et al. (2018) and Ferreira et al. (2022). A comprehensive and integrated compilation of this literature-based information obtained through the SLR conducted was performed, and further completed with some additional contribution by the authors, as presented in Table 2.

In collaboration, there are some main key words that are more or less closely implied, such as interaction, interoperation, integration, distribution, decentralization, networking, which may further imply an increased complexity, arriving not just from the great amount and diversity of data/information shared, but also by the unpredictable interrelations and interchanges of this data/information, through a more or less widened network of collaborating entities, for instance, machines in an extended manufacturing environment context.

This complexity will, in the limit, conduct to the necessity of using approaches and tools for supporting decision support, namely based on chaos and complexity management, game theory, group decision-making, organizational semiotics, and learning organization principles, along with machine/deep learning, among other intelligent management and DM paradigms and methods, including more autonomous or automatic ones, for instance, based on MAS which is typically used in M-M collaboration.

Table 2. Synthesis of main collaborative manufacturing and management dimensions in Industry 4.0 (adapted from: Varela et al. (2023).

Collaboration in I4.0	Dimensions	Characteristics
Paradigms, methodologies, and business strategies	D1.1: Open design and innovation strategies, smart product design, concurrent design theory, and others	Designing smart products, and advanced materials, and/or production systems, based on concurrent and collaborative engineering, systems and organizations, and design theory
	D1.2: Metatheory, and Learning Organizations, organizational semiotics and norms, pragmatics, and others	Metatheory, formal theories and formalisms, learning organization, organizational learning and semiotics, and norms in collaborative processes and practices
	D1.3: New business models, circular economy, finance, and risk management/ assessment, global resources management	New business and organization models and technology transfer to companies, attending the principles underlying the circular economy, along with finance and risk management strategies and approaches
	D1.4: Organizational change and transformation	Organizational change and transformation, regarding stakeholders functions, and employees, along with their competencies, culture management, and sustainable needs
	D1.5: Advanced energy collection, storage, and management strategies, decarbonization, and other sustainability policies	Advanced energy collection, and storage management, and decarbonization as additional issues to reach sustainable collaborative processes and practices in and between companies.
Models, frameworks, architectures, approaches, tools, and implementation technology	D2.1: Servitization, digitalization, decentralization, parallelism	Integration, distributivity, virtuality, agility, servitization, digitalization, and decentralization, and parallelism as major issues in and for collaborative engineering

Table 2. contd. ...

Table 2. contd. ...

Collaboration in I4.0	Dimensions	Characteristics
	D2.2: C[P]PS, smart factories, the factory of the future, advanced robotics, and automation, and others	Cyber Physical [Production] Systems, and smart factories, based on intelligent sensing systems, open systems, networked, and distributed manufacturing systems; and further urban production systems, virtual organizations, open systems, along with learning organizations, and other technologies for advanced manufacturing and management
	D2.3: [I]IoT, and smart/ sensing, and collaborative networks, and others	[Industrial] Internet of Things, smart, ubiquitous, cloud-based, large and complex networks and technology aiming at high digitalization levels
	D2.4: Horizontal and vertical integration	Horizontal and vertical integration among stakeholders, including partners, factories, suppliers, customers, and diverse businesses
	D2.5: Cybersecurity and cyber control	Cybersecurity, and cyber control for establishing proper connections, and communications, in data and information transferring, sharing, and processing
	D2.6: Cloud and grid computing, manufacturing, and management, and others	Cloud based computing, manufacturing and management for fully proving enhanced flexibility and suitability in collaborative processes and practices
	D2.7: Additive manufacturing	Additive manufacturing or 3D printing as additional key enabling technology and principles for promoting collaborative processes and practices between entities and stakeholders

Table 2. contd. ...

Table 2. contd. ...

Collaboration in I4.0	Dimensions	Characteristics
	D2.8: Exponential technology, and supercomputing or high performance computing	Advanced production technology, systems, and processes, high performance computing, exponential, and disruptive technologies, e.g., automation and robotics, autonomous and collaborative robots, advanced mechatronics, and micro and nano manufacturing as further key enablers of collaboration (H-M and M-M)
	D2.9: Advanced interfaces, virtual, augmented and mixed reality, digital twin, and others	Advanced interfaces, virtual, augmented, and mixed reality, and digital twin, promoting and enhancing collaboration in manufacturing and management
	D2.10: Artificial Intelligence approaches, techniques, and methods, and technology	[Applied] Artificial Intelligence, machine learning and deep learning, pattern recognition, blockchain, and other technologies and methodologies for enabling and enhancing collaboration
	D2.11: Business intelligence, data science, and big data analytics	Business intelligence, big data, and data analytics in the data science domain as important issues to enable and promote collaboration
	D2.12: Advances SCM, ERP, and MES, CIM, and others	Advanced, integrated and intelligent supply network management, project, and business management, collaborative and integrated manufacturing, management, quality, maintenance, performance measures: sustainability, organizational and machine robustness, scalability systems, along with other advanced EIS, such as ERPs, MES, and SCM

Table 2. contd. ...

Table 2. contd. ...

Collaboration in I4.0	Dimensions	Characteristics
	D2.13: Advanced DSS, DB, KB, and Semantic web, and others	Advanced/ integrated decision support systems (DSS), and databases (DB), knowledge engineering and knowledge bases (KB), automatic data acquisition, and semantic web for enhancing collaboration
	D2.14: Advanced Simulation, game theory, technology, chaos and complexity management, and others	Advanced simulation, and other management systems based on game theory, and other advanced and integrated optimization approaches, e.g., for chaos and complexity management, based on distributed, collaborative and real-time management principles

Moreover, collaboration is also fostered or can be enhanced by the use of other recent technologies, for instance for improving information and knowledge sharing capabilities, namely through the use of a widened set of AI-based approaches and platforms that enable 'servitization', emergence, social communities, and networks, along with varying kind of internet based paradigms, protocols and technologies, to be used among collaborating organizations. These organizations may collaborate through extended supply networks, for establishing interconnections between business partners, in virtual, agile and distributed enterprises, supported by entrepreneurship philosophies, along with advanced ICT, and exponential technologies, e.g., High Performance Computers (HPC), along with other technologies and principles underlying the I4.0, for instance, based on parallel tasks programming.

Therefore, the I4.0 and underlying technology can enhance or promote collaboration, by enabling full digitalization of everything, vertical and horizontal integration or entities, and point-to-point or end-to-end access and communication, along with agile IT technologies, platforms, and services for improved interconnections and information, knowledge and resources sharing and co-working or co-creation (https://twitter.com/mikequindazzi/status/829993822008532992).

Thus, the so called smart or intelligent industry will further be based on different collaboration levels, varying from process, and technology to the full organizational and inter personal levels, not just in terms of intra- but also inter-organizations and among stakeholders worldwide. These will be spread through a widened range of globally distributed points, of not just physical but also virtually distributed and complex networks of entities, including not only factories but varying, heterogeneous, and extended sets of inter-players, including machines, and tools. Further these complex networks will include suppliers, and customers, through extended supply networks, as is also reinforced by the determination of the so-called "Smart Industry 4.0 readiness index", through which it is considered that "Inter- and Intra-Company Collaboration" is a fundamental condition in the scope of Organizations' structure and management (Pfaff and Hasan, 2011).

Moreover, also in Ustundag and Cevikcan (2017), the authors mention that the current digital era differs from the others by not just providing changes in main business processes but also by revealing concepts of smart and connected products, along with service-driven business models that enable to increase collaboration in the production network through consistent data availability, along with the use of exponential technologies for offering multiple benefits, being the enhanced productivity just a starting point.

In Srivastava (2008), the collective intelligence concept and its importance for the so-called new corporate governance is further explored. According to the authors, relevant new concepts, technological aids and events are a part or contribute in framing their proposed new corporate governance model based on Collective Intelligence and Knowledge Management. For instance, based on amplified intelligence technology, acquired information, information society, collective reflection, collective DM, and extended organization.

Collaboration M&M processes and practices, along with supporting collaborative networks and platforms, decision systems, web applications and services, namely, for supporting engineering and production management, are of utmost importance, and continue growing nowadays, in the context of the I4.0 (Putnik and Putnik, 2019; Putnik et al., 2021a,b,c; Ferreira et al., 2022; Manupati et al., 2022; Varela et al., 2022).

The production systems that currently still prevail in the vast majority of industrial companies, internationally and worldwide, are traditional production systems, with little or no degree of either automation or some form of 'intelligence' or 'smart' characteristic, thus being far away from the envisioned I4.0 scope.

Moreover, this reality is even more problematic due to the absence of a truly collaborative oriented culture in companies (Putnik and Putnik, 2019; Putnik et al., 2021a,b,c; Ferreira et al., 2022; Manupati et al., 2022; Varela et al., 2022), in which the technological issues and concerns are overwhelming and not effectively supported by a real mind shift regarding the consciousness about the fundamental importance of the integration of collaboration processes and practices, along side with business and organizational change in I4.0.

Therefore, production systems will have to undergo profound changes, in a medium term perspective, in order to guarantee collaboration, as it is a major issue to enable reaching real sustainability in the light of current global market and production requirements and of the ongoing and eminent development, and absolutely necessary for its evolution and prosperity in I4.0 and further. Thus, in the context of the current I4.0 industrial revolution, companies are already starting to be forced to restructure their production systems and the corresponding administration, production and management paradigms, in order to evolve, in a fast way, towards the concept of CP[P] based on the evolution of current production systems (by refurbishing, etc.) and/or based on the creation of new production units, and open or flexible systems, completely oriented to the C[P]PS paradigms in the I4.0 concept and the underlying promoting approaches and technologies (Nikolakis et al., 2019; Ferreira et al., 2022; Manupati et al., 2022).

In this sense, a fundamental contribution to this end consists on the use of collaborative M&M processes and practices, aligned not just with current main I4.0 dimensions but also with some additional and more specific ones that are required from the Collaborative Manufacturing and Management scope, which constitutes a fundamental priority in order to further enable the majority of existing companies to face this current need of change and upgrade, not just technologically but through the essential support that arises from groups of specialized professionals that jointly gather fundamental and new knowledge in the two underlying main interchangeable scientific domains and practices (I4.0 and Collaborative Engineering).

5. Discussion on Collaborative Manufacturing and Management in the Industry 4.0

According to the main literature contributions analyzed, it is noticeable that in the I4.0 some main issues or dimensions underlying CollM&M arise from the AI domain, along with other technologies, which include [I]IoT, digitalization, servitization, flexibilization, dynamism, complexity, distribution, parallelism, and automation, along with agility, networking and virtualization.

I4.0 is dominated by a wide range of technologies and methodologies, which requires a human aspect for its full or proper operation, in addition to the industrial production engineering aspect, which underpins the scientific knowledge of the methods, techniques, tools and methodologies of development and management of CPS and underlying production processes. The aspects of engineering, industrial management, economics, and human engineering deal with a wide range of skills in order to allow success in business and in its management, with a view to innovation, sustainability, and entrepreneurship. It is thus expected that joint I4.0 and collaboration oriented approaches and practices will permit to conceive and develop suitable and innovative models, methodologies, approaches and systems, which include the product design phase, the production process, the distribution of the final product across global markets, and also taking into account reengineering, reverse engineering, along with other related engineering and manufacturing processes, along widened extended business and supply networks.

6. Conclusion

In this chapter a general view over different approaches and application scenarios of the most relevant collaboration and related terms were explored, along with an overview of the main topics that have been put forward worldwide about collaboration within a set of 19 main Industry 4.0 (I4.0) dimensions identified. The main collaboration and I4.0 topics were jointly studied based on a systematic literature review (SLR). Through the literature analysis it was possible to recognize that CollM&M is correlated with approaches and technology underlying the I4.0, which can further promote and support it.

It was possible to realize that the widened port folio of contributions in the literature about CollM&M-I4.0 can be grouped into seven main clusters, about: (1) learning approaches, including machine/deep learning models and systems, among other approaches and supporting technology, namely, for supporting human-centered CollM&M decision-making processes; (2) Design and performance evaluation of life-cycle assessment strategies, environments, through diverse kind of business models, and ecosystem services for reaching sustainable development of companies; (3) Different types of decision and communication support technology, including platforms, frameworks, and simulation models and tools; (4) Advanced architectures, methodologies, and models based on diverse kinds of recent approaches, namely, based on cloud, game theory, and chaos and complexity management approaches and systems; (5) AI-based approaches, based on different procedures for enabling and supporting CPPS, including multi-agents, pattern recognition, along with high-performance computing and computer vision technology, among others; (6) Data science- centered contributions, and further based on IoT, besides the integration of hard and software tools for enabling flexible and interactive decision support, for instance, through digital twins, and big data analytics; and (7) Other CollM&M approaches, based on a widened range of algorithms, methods and tools, namely, for supporting group decision- making, and patterns and performance analysis.

Future research should thus focus on the development of new approaches and technologies to enable human-machine collaboration, along with the exploration of further machine-machine collaboration, in order to evaluate its implementation in diverse organizations and industrial sectors, namely for supporting manufacturing and management practices.

References

Alves, C., Putnik, G. D., and Varela, M. L. R. How Environment Dynamics Affects Production Scheduling: Requirements for Development of CPPS Models. *FME Trans.*, 2021, *49*, 827–834. https://doi.org/10.6036/9997.

Arrais-Castro, A., Varela, M. L. R., Putnik, G., Ribeiro, R., and Dargam, F. Collaborative negotiation platform using a dynamic multi-criteria decision model. *Int. J. Decis. Support Syst. Technol.*, 2015, *7*, 1–14. https://doi.org/10.4018/ijdsst.2015010101.

Bechtold, J., and Lauenstein, C. (2014). Digitizing Manufacturing: Ready Set Go. *Capgemini. https://www. de. capgemini-consulting. com/resource-file-access/resource/pdf/digitizing-manufacturing_0. pdf. Zugegriffen am*, *7*, 2016.

Canadas, N., Machado, J., Soares, F., Barros, C., and Varela, L. Simulation of cyber physical systems behaviour using timed plant models. *Mechatronics*, 2018, *54*, 175–185. https://doi.org/10.1016/j.mechatronics.2017.10.009.

Chen, D. (2005). Enterprise-control system integration—an international standard. *International Journal of Production Research*, *43*(20), 4335–4357.

Chen, X., and Zhou, J. (2021). The complexity analysis and chaos control in omni-channel supply chain with consumer migration and advertising cost sharing. *Chaos, Solitons, & Fractals*, 146, 110884.

Costa, J., Ávila, P., Bastos, J., Pinto, and Ferreira, L. A New Simple, Flexible, and Low-Cost Machine Monitoring System. *DYNA*, 2021, *96*, 640–646. https://doi.org/10.6036/10075.

Deloitte Consulting, L. L. P., and by Deloitte, B. *Global Human Capital Trends 2014: Engaging the 21st-century Workforce*. 2014b.

Deloitte. *Industry 4.0: Challenges and Solutions for the Digital Transformation and Use of Exponential Technologies*. 2014a.

Eijnatten, F. V., and Putnik, G. D. Chaos, complexity, learning, and the learning organization: Towards a chaordic enterprise. *Learn. Organ.*, 2004, *11*, 418–429.

Ferreira, L., Punik, G., Varela, L., Manupati, V. K., Lopes, N., Cunha, M., Alves, C., and Castro, H. A Framework for Collaborative Practices Platforms for Humans and Machines in Industry 4.0 oriented Smart and Sustainable Manufacturing Environments. *In*: *Smart and Sustainable Manufacturing Systems for Industry 4.0*; Manupati, V.K., Goran, D.P., Rocha, M.L. (Eds.). CRC Press, Taylor & Francis Group: Boca Raton, FL, USA, 2022; in press.

Finance, A. T. C. C. Industry 4.0 Challenges and solutions for the digital transformation and use of exponential technologies. *Finance. Audit. Tax Consult. Corp.*, Zur. Swiss, 2015, 1–12.

Fonseca, L. The EFQM 2020 model. A theoretical and critical review. *Total Qual. Manag. Bus. Excell.* AHEAD-OF-PRINT, pp. 1–28, https://doi.org/10.1080/14783363.2021.1915121.

Fonseca, L. Amaral, A. and Oliveira, J. Quality 4.0: The EFQM 2020 Model and Industry 4.0 Relationships and Implications. *Sustainability*, 2021, *13*, 3107. https://doi.org/10.3390/su13063107.

Hankel, M., and Rexroth, B. The reference architectural model industrie 4.0 (Rami 4.0). ZVEI, 2015, 2(2), 4–9.

Ibarra, D., Ganzarain, J., and Igartua, J. I. Business model innovation through Industry 4.0: A review. *Procedia Manufacturing, 2018, 22,* 4–10.

Kagermann, H., Helbig, J., Hellinger, A., and Wahlster, W. *Recommendations for Implementing the Strategic Initiative INDUSTRIE 4.0: Securing the future of German manufacturing industry*; final report of the Industrie 4.0 Working Group. Forschungsunion. 2013.

Lee, J., Bagheri, B., and Kao, H. A. A cyber-physical systems architecture for industry 4.0-based manufacturing systems. *Manuf. Lett.*, 2015, *3*, 18–23.

Li, J. Q., Yu, F. R., Deng, G., Luo, C., Ming, Z., and Yan, Q. Industrial internet: A survey on the enabling technologies, applications, and challenges. *IEEE Commun. Surv. Tutor.*, 2017, *19*, 1504–1526.

Liao, Y. Loures, E. D. F. R., and Deschamps, F. Industrial Internet of Things: A systematic literature review and insights. *IEEE Internet Things J.*, 2018, *5*, 4515–4525.

Manyika, J., Chui, M., Brown, B., Bughin, J., Dobbs, R., Roxburgh, C., and Hung Byers, A. *Big Data: The Next Frontier for Innovation, Competition, and Productivity*; McKinsey Global Institute: Washington, DC, USA, 2011.

Nikolakis, N., Maratos, V., and Makris, S. A cyber physical system (CPS) approach for safe human-robot collaboration in a shared workplace. *Robotics and Computer-Integrated Manufacturing*, 2019, *56*, 233–243.

Prades, L., Romero, F., Estruch, A., García-Domínguez, A., and Serrano, J. Defining a methodology to design and implement business process models in BPMN according to the standard ANSI/ISA-95 in a manufacturing enterprise. *Procedia Eng.*, 2013, *63*, 115–122.

Putnik, G., and Ávila, P. Manufacturing System and Enterprise Management for Industry 4.0 (Editorial). *FME Trans.*, 2021, *49*, 769–772. https://doi.org/10.5937/fme2104769P.

Putnik, G. D., Putnik, Z, ; Shah, V., Varela, L., Ferreira, L., Castro, H., Catia, A., and Pinheiro, P. Collaborative Engineering: A Review of Organisational Forms for Implementation and Operation. *In*: *IOP Conference Series: Materials Science and Engineering*; IOP Publishing: Bristol, UK, 2021; Volume 1174, p. 012028.

Putnik, G.D., Putnik, Z., Shah, V., Varela, L., Ferreira, L., Castro, H., Catia, A., and Pinheiro, P. Collaborative Engineering definition: Distinguishing it from Concurrent Engineering through the complexity and semiotics lenses. *In*: *IOP Conference Series: Materials Science and Engineering*; IOP Publishing: Bristol, UK, 2021; Volume 1174, p. 012027.

Putnik, G. D., Škulj, G., Vrabič, R., Varela, M. L. R., and Butala, P. Simulation Study of Large Production Network Robustness in Uncertain Environment. *CIRP Ann.—Manuf. Technol.*, 2015, *64*, 439–442. https://doi.org/10.1016/j.cirp.2015.04.118.

Putnik. G. D., Pabba, S. K., Manupati, V. K., Varela, M. L. R., and Ferreira, F. Semi-Double-loop machine learning based CPS approach for predictive maintenance in manufacturing systems based on machine status indications. *CIRP Ann.—Manuf. Technol.*, 2021, *70*, 365–368. ISSN 0007-8506. https://doi.org/10.1016/j.cirp.2021.04.046.

Reddy, M. S., Ratnam, C., Agrawal, R., Varela, M. L. R., Sharma, I., and Manupati, V. K. Investigation of reconfiguration effect on makespan with social network method for flexible job shop scheduling problem. *Comput. Ind. Eng.*, 2017, *110*, 231–241. https://doi.org/10.1016/j.cie.2017.06.014.

Rodič, B. Industry 4.0 and the new simulation modeling paradigm. *Organizacija*, 2017, *50*, 193–207.

Shah, V., and Putnik, G. D. Machine learning based manufacturing control system for intelligent Cyber-Physical Systems. *FME Trans.*, 2019, *47*, 802–809.

Varela, M. L. R., Aparício, J. N., and Silva, S. C. A Web-based application for manufacturing scheduling. *In*: *IASTED International Conference on Intelligent Systems and Control*; Salzburg, Austria, 2003; pp. 400–405, 25th to 27th of June 2003, ISBN: 0889863555.

Varela, M. L. R., and Carmo Silva, S. An ontology for a model of manufacturing scheduling problems to be solved on the web. *In*: *International Conference on Information Technology for Balanced Automation Systems*; Springer: Boston, MA, USA, 2008; pp. 197–204.

Varela, M. L. R., Putnik, G. D., Manupati, V. K., Rajyalakshmi, G., Trojanowska, J., and Machado, J. Collaborative manufacturing based on cloud, and on other I4.0 oriented principles and technologies: A systematic literature review and reflections. Manag. *Prod. Eng. Rev.*, 2018, *9*, 90–99. https://doi.org/10.24425/119538.

Varela, M. L. R., Putnik, G. D., Manupati, V. K., Rajyalakshmi, G., Trojanowska, J., and Machado, J. Integrated Process Planning and Scheduling in Networked Manufacturing Systems for I4.0: A review and framework proposal. *Wirel. Netw.*, 2019, *27*, 1587–1599.

Varela, M. L. R., and Ribeiro, R. A. Distributed manufacturing scheduling based on a dynamic multi-criteria decision model. *In*: *Recent Developments and New Directions in Soft Computing*; Springer: Cham, Switzerland, 2014; pp. 81–93.

Xu, L. D., Xu, E. L., and Li, L. Industry 4.0: State of the art and future trends. *Int. J. Prod. Res.*, 2018, *56*, 2941–2962.

4

Collaborative Manufacturing Management Meta-model

Nowadays, global resources management intersects with collaboration and Industry 4.0 paradigms, namely, for collaboratively managing cyber-physical systems. Only organizations that cooperate with their business partners, along with their suppliers and remaining stakeholders, including their clients, will be able to permit and promote the much-needed endowing of agility, effectiveness, and efficiency in their management processes. For that, suitable decision-making paradigms, along with underlying approaches, will be needed, in order to properly fulfil current companies' decision requirements and practices. The main purpose of this chapter is to show that this can be achieved by applying combined global resources management paradigms and approaches, to reach collaboration further supported by recent technology made available through Industry 4.0. In doing so, the interaction of companies and stakeholders, supported by appropriate networks, along with varying kinds of other communication and problem-solving technology, will enable them to promote and reinforce interoperation to reach the best-suited management decisions, by considering each one's objectives and priorities, along with common goals. To this end, in this chapter, a systematic literature review methodology is mentioned to synthetize the main contributions about the relation of these domains. The study carried out and the results obtained permitted us to realize that dynamic, integrated, distributed, parallel, intelligent, predictive, and real-time-based decision paradigms are of utmost importance currently, but are still just scarcely being combined, which is suggested though its encompassing through a proposed collaborative management framework that is recommended to be applied, either in industry or academia, to improve global resources management processes and practices.

1. Introduction

Collaborative Management (CollManag) requires the application of management processes and approaches of a more or less widened set of companies and stakeholders that interact for solving some shared problem, usually intending to reach some common goal, besides their own objectives and priorities.

Collaborative networks (CN), and global or group decision-making approaches (GDMA) are fundamental for enabling and promoting the interaction and sharing of knowledge among two or more collaborating entities (Putnik et al., 2021a,b; Varela et al., 2022a,b,c). Moreover, independently of sharing or not having the same goal, and/or resources, usually, interplaying entities do fall into some kind of business environment, for instance, in the context of distributed or extended manufacturing systems (EMS) or agile/virtual enterprises (A/VE) (Lou et al., 2010; Vieira et al., 2012; Putnik and Cruz-Cunha, 2005).

In the current complex and turbulent manufacturing environments (Eijnatten and Putnik, 2004), such as EMS or A/VE, it is fundamental to make use of CN and GDMA, in order to fulfil the requisites imposed by Industry 4.0 (I4.0) (Putnik and Ferreira, 2019), and to solve the shared management problems, for instance, related to manufacturing planning and scheduling, occurring either in more traditional or in EMS or A/VE manufacturing environments or in cyber-physical production systems (CP[P]S) (Low et al., 2013; Guo et al., 2015; Canadas et al., 2018; Alves et al., 2021), thus, usually requiring some combination of management paradigms and approaches (P&A) for solving complex and distributed manufacturing scheduling (DMS) problems.

A DMS problem (Vieira et al., 2012; Alves, Putnik, and Varela, 2021) is one typical example of the need for using CN and GDMA for solving the scheduling problem among a set of participating companies, which may or may not further share manufacturing resources and be geographically dispersed, tending to be quite complex combinatorial optimization problems (Varela et al., 2022c; Varela and Ribeiro, 2014).

The use of GDMA is fundamental to enable the resolution of DMS problems, among others, occurring in the scope of CollManag, based on proper approaches, methods, and techniques, along with the use of appropriate communication networks among the set of interacting entities or companies.

Besides distributed scheduling, other important issues do occur in the scope of CollManag, namely, related to dynamically changing production conditions and customers' order requisites, in real-time, along with the need for integrating varying kind of other management issues, besides scheduling ones, related, for instance, to maintenance management, among others, that also influence the whole global management process and increase its complexity (Putnik and Cruz-Cunha; 2006; Putnik et al., 2021).

Thus, it becomes imperative to make use of appropriate decision support (DS) approaches and tools, which enable dynamic and agile DS, namely, through the use of multi-criteria decision- making (MCDM) methods and models (Varela and Ribeiro, 2014), along with intelligent and/or predictive DS algorithms and systems (Azevedo et al., 2021; Azevedo et al., 2022), besides other approaches and technologies, for instance, to permit parallel programming (Lopes et al., 2022). This last one can further benefit from different kinds of Industry 4.0 (I4.0) technology, namely from the use of high-performance computing (HPC) (Lopes et al., 2022).

Thus, this chapter intends to contribute to the synthesis of the main research and findings about CollManag-P&A, during the last decade, by highlighting the importance of supporting I4.0 technology, and to enable answering the following research question: "What are the main decision support paradigms and approaches underlying collaborative management in the current digitalization era to promote a sustainable development of companies?"

Moreover, both CollManag-P&A and I4.0 together, can be seen as collaborative decision-making processes and practices that are currently fundamental to enable and promote decision-making in and between companies and their stakeholders, namely, in the context of CPPS, and to permit to reach the endowing of agility, effectiveness, and efficiency of their management processes.

The synthesis and detailed analysis performed in this chapter based on the application of a systematic literature review (SLR) enabled us to identify a varying set of CollManag-P&A that enable supporting companies to properly address their daily management decision-making processes.

Moreover, it was possible to identify a set of main CollManag paradigms that were encompassed in a proposed collaborative management meta-model, in the I4.0 context. The proposed meta-model is, thus, intended to enable solving more or less complex CollManag problems, namely in EME or A/VE, or in the context of cyber-physical production systems (CPPS) (Putnik and Ferreira, 2019; Putnik et al., 2021), which play a very important role nowadays in the I4.0 era. This proposed meta-model integrates dynamic, distributed, integrated, intelligent, predictive, parallel, and real-time-based approaches, for fulfiling the requirements underlying the resolution of the CollManag problems that may occur nowadays in different kinds of manufacturing environments.

These manufacturing environments may vary from more classical or centralized manufacturing environments up to fully distributed and decentralized ones. Moreover, the proposed CM framework is a novel contribution, and as to our knowledge, there is not yet any such kind of contribution available in the literature. Thus, some more specific ones are made, regarding the resolution of some kinds of problems occurring in a more or less concrete manufacturing environment or application scenario are usually being explored, and/or based on a reduced combination of management paradigms and underlying approaches. Instead, by considering our proposed CollManag meta-model, different kinds of management P&A can be combined, along with varying types of underlying methods/algorithms, and corresponding problem-solving tools or platforms, for solving a CollManag problem. Therefore, different combinations of appropriate methods and techniques, varying from applying pure mathematical optimization methods to the use of diverse types of metaheuristics, among other artificial intelligence (AI) approaches, e.g., machine learning or multi-agent systems (MAS), just to mention a few, may be applied for solving the CollManag problems, among others (Varela et al., 2022c; Vieira et al., 2012; Azevedo et al., 2021; Azevedo et al., 2022; Alves et al., 2019).

Summarizing, this chapter aims at briefly presenting and discussing collaborative manufacturing management approaches in the current I4.0 context, based on results obtained through an SLR. The reviewed approaches are related to dynamic, distributed, parallel, predictive, and real-time based paradigms, and the evolution of its application in manufacturing management decision- making over the last decade (2011–2021). The results obtained demonstrated the increasing importance that these management approaches are assuming today, either used independently or combined in the current digitalization era. The study also revealed the supporting technologies that are also increasingly being used, associated with those management approaches.

I4.0 represents the current trend of automation technologies in the manufacturing industry, and it mainly includes enabling technologies such as the CPS, Internet of Things (IoT), cloud computing, AI, blockchain, industrial information integration, and other related technologies (Li, 2018).

According to a common definition of manufacturing management, regarding one of its main functions related to manufacturing scheduling, consists on the allocation, over the time, of jobs to machines, within a short temporal horizon, and according to a specific criterion, such as cost or tardiness (Framinan and Ruiz, 2010). Manufacturing scheduling systems is a research subject extensively explored in the literature, either documented in many quantitative and qualitative studies (Framinan and Ruiz, 2010) or covered by deep review and research frameworks (Gahm et al., 2016). The MS research offered in literature cover different approaches, ranging from framework concepts as sustainability (Akbar and Irohara, 2018) or energy-efficiency (Gahm et al., 2016), to supporting technologies such as cloud (Liu et al., 2019), machine learning (Dogan and Birant, 2021), etc., just to mention some.

Manufacturing management, and scheduling assume a fundamental relevance in I4.0, as today manufacturing scheduling is performed within a smart manufacturing environment, with a special focus on paradigms and approaches related to dynamic, distributed, parallel, predictive, and real-time based methods and tools. However, literature does not provide a comprehensive and integrated view of these manufacturing management paradigms in the actual context of I4.0. The authors conducted an SLR in order to cover this gap, and to help understand the importance that the different kinds of the identified manufacturing management paradigms assume, namely, in the scope of manufacturing scheduling decision-making in the current digitalization era.

Literature produced during the 10-year period between 2011 and 2021 about the current I4.0 dynamic, distributed or decentralized, parallel, predictive or intelligent and real-time based models, methods, and tools was reviewed. By doing so, this work has a twofold contribution: on one hand, it offered a comprehensive overview on the increasing importance that these scheduling approaches are assuming today, either used independently or combined, while on the other hand, it simultaneously revealed the supporting technologies that are also increasingly being used, associated with those manufacturing management, for instance, scheduling approaches.

The remainder of the chapter is structured as follows: in the next section, a brief description of a proposed collaborative management meta-model is provided, along with the underlying paradigms, focusing on dynamic, distributed, parallel, integrated, intelligent/ predictive, and real-time based management. In the Section 3, main collaborative management paradigms and approaches, and their application in the current digital era are briefly described. Further, in Section 4, the contextualization of the proposed collaborative management meta-model within the state of the art, and future research and application directions are discussed. Finally, in Section 5, the main conclusions reached are presented, along with some further research questions.

2. Collaborative Manufacturing Management Meta-model and Definition

Adaptable and Intelligent manufacturing systems (IMS) are increasingly required nowadays to meet the increasing requirements of customers and factories aiming at high quality levels and responsiveness. In the current I4.0 context, enabled by the (CPPS) and rooted on IMS (Monostori, 2014), MS is a decision-making process that has been significantly improved and automated (Varela et al., 2021). Although, the human continues to assume a central role or importance in appropriately conducting manufacturing management decision- making, under different kinds of management paradigms (Varela et al., 2022; Bolton, 2008). The new manufacturing structures will induce changes in the way production planning, programming, and control is carried out, and a smart approach to solve production scheduling problems has been proposed (Monostori, 2014).

In this section, a meta-model about CollManag will be presented, which hereafter will be named as collaborative management (CM), in the I4.0 context, and which resulted from the SLR carried out, in addition to the co-authors' own knowledge in the focused scientific domain, as can be seen, for instance, in (Leitão 2008; Hsu and Yang, 2016; Zhang et al., 2019; Rohaninejad et al., 2021; Nof and Grant, 1991; Ghaleb et al., 2020).

The proposed CM framework is considered to be of utmost importance currently in the I4.0 era, as it includes a set of the following six main identified management paradigms from the literature: integrated, dynamic, intelligent/predictive, distributed, parallel, and real-time management (Figure 1).

Next, the main CollManag paradigms in I4.0 will be briefly described.

Dynamic Management

As stated in (Villalonga et al., 2021), in the currently digitalization era, dynamic production management is crucial, for enabling to reach appropriate decisions, which is a big challenge, namely, in the context of CP[P]S, for which the use of automated decision-making approaches, along with digital twin, and fuzzy inference systems, can be highly beneficial, being executed on the due course of the production. The authors propose a framework for decentralized and integrated decision-making for re-scheduling of a CPPS, which they mention was successfully validated in an I4.0 pilot line of assembly process.

In Ebufegha (2021) is also presented a model for dynamic scheduling in smart manufacturing systems (SMS) and simulating its operation. SMS are characterized as highly modular and flexible systems, with every physical resource being able to exchange information with each other

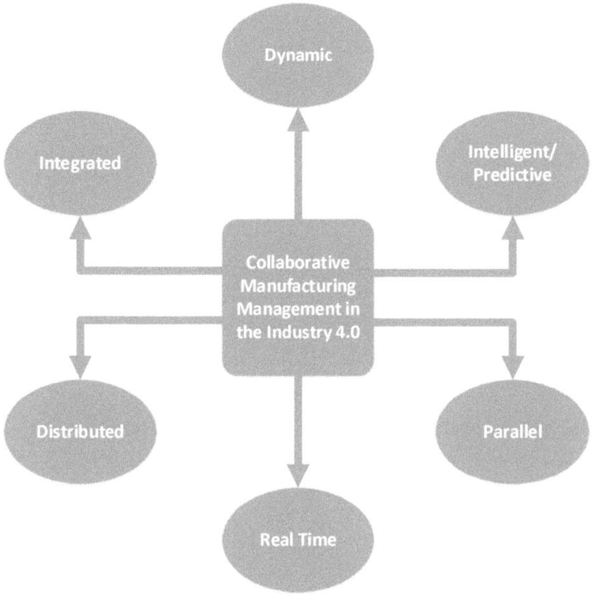

Figure 1. Collaborative management meta-model (adapted from Varela et al., 2023).

over an industrial network and self-organize to schedule operations in real-time. According to the authors, this can potentially reduce orders' completion times and increase the average machine utilization. Additional work related to dynamic scheduling is presented in (Saboor et al., 2019; Hofer et al., 2020; Tan et al., 2019; Ferreirinha et al., 2019).

Collaborative management paradigm based on dynamism at different levels: data/information, methods, procedures, tools, and platforms to support different management decision-making processes, namely in:

- **Collaborative Dynamic Decision Making: A Case Study from B2B Supplier Selection** (Campanella et al., 2012; *LNBIP, Springer*).
- **Dynamic MCDM with Future Knowledge for Supplier Selection** (Jassbi et al., 2014; *JDS, Taylor & Francis*).
- **Distributed Manufacturing Scheduling based on a Dynamic Multi-criteria Decision Model** (Varela and Ribeiro, 2014; *SFSC, Springer*).
- **Collaborative Negotiation Platform using a Dynamic Multi-criteria Decision Model** (Arrais-Castro et al., 2015a; *IJTSST, IGI Global*).
- **Spatial-temporal Business Partnership Selection in Uncertain Environments** (Arrais-Castro et al., 2015b; *FME Transactions, Belgrade Univ.*).

- **Collaborative Framework for Virtual Organisation Synthesis based on a Dynamic Multi-criteria Decision Model** (Arrais-Castro et al., 2018; *IJCIM, Taylor & Francis*).
- **A Collaborative Multiplicative Holt-Winters Forecasting Approach with Dynamic Fuzzy-Level Component** (Kays et al., 2018; *Appl. Sci., MDPI*).
- **Normalization Techniques for Collaborative Networks** (Vafaei et al., 2019, *Kybernetes; Emerald Group Pub. Ltd.*).
- **Decision Support Tool for Dynamic Scheduling, and an Industry 4.0 Oriented Tool for Supporting Dynamic Selection of Dispatching Rules based on Kano Model Satisfaction Scheduling** (Ferreirinha et al., 2019a,b; *FME Transactions, Belgrade Univ.*).
- **How Environment Dynamics Affects Production Scheduling: Requirements for Development of CPPS Models** (Alves et al., 2021; *FME Transactions, Belgrade Univ.*).

Integrated management

Collaborative management paradigm based on integration at different levels: data/ information, methods, procedures, tools, decision support systems, and platforms, including the integration of diverse management functions, namely, in:

- **Web-based Technologies Integration for Distributed Manufacturing Scheduling in a Virtual Enterprise** (Varela et al. 2012; *IJWP, IGI Global*)
- **Technologies Integration for Distributed Manufacturing Scheduling in a Virtual Enterprise** (Vieira, Varela, and Putnik, 2012; *CCIS, Springer*)
- **A Web-Based DSS for Supply Chain Operations Management towards an Integrated Framework** (Carvalho et al., 2014; *LNBIP, Springer*)
- **An Integer Programming Approach for Balancing and Scheduling in Extended Manufacturing Environment** (Kays et al., 2015; *IEEEXplore*)
- **Scheduling and Batching in Multi-site Flexible Flow Shop Environments** (Santos et al., 2015; *IEEEXplore*)
- **Integrated Platform for Real-time Control and Production and Productivity Monitoring and Analysis** (Vieira et al., 2016; *RRPMOM, Elsevier*).
- **Collaborative Manufacturing based on Cloud, and on Other I4.0 Oriented Principles and Technologies: A Systematic Literature**

Review and Reflections (Varela et al., 2018b; *MPER, Polish Acad Sciences*).

- **Hybrid System for Simultaneous Job Shop Scheduling and Layout Optimization based on Multi-agents and Genetic Algorithm** (Alves et al., 2018; *AISC, Springer*).
- **Integrated Process Planning and Scheduling in Networked Manufacturing Systems for Industry 4.0: A Review and Framework Proposal** (Varela et al., 2021; *Wireless Networks, Springer*).
- **Group Decision-making Approach for Ranking and Selecting Maintenance Tasks for being Jointly Scheduled with Production Orders** (Varela et al., accepted; *IJQR, Univ. Montenegro*).

Distributed management

Manufacturing management, namely scheduling, in distributed manufacturing environments is a particularly complex combinatorial problem, and even further when considering dynamic environments (Leitão 2008). A holonic approach for distributed scheduling in a dynamic manufacturing environment is presented in (Leitão 2008), where the scheduling functions are distributed by several entities. The authors propose a scheduling and control approach that aims at achieving fast and dynamic re-scheduling by using a scheduling mechanism that evolves dynamically to combine centralized and distributed strategies, improving its responsiveness to emergence, instead of the complex and optimized scheduling algorithms found in traditional approaches. Other interesting contributions are being put forward, with a special focus on the current I4.0 requirements and increasingly complex manufacturing conditions (Liu et al., 2019; Varela et al., 2021; Hsu and Yang, 2016; Zhang et al., 2019; Jiang et al., 2021; D'Aniello et al., 2021; Sousa and Oliveira, 2020; Lohmer and Lasch, 2021).

Collaborative management paradigm that is based on distributed management models, processes, approaches, tools, systems, and platforms, namely for supporting distributed scheduling, among other management functions, namely, in:

- **A Web Interface for Accessing Scheduling Methods in a Distributed KB** (Varela et al., 2004; *IFIP AICT, Kluwer Acad. Publ.*).
- **Definition of a Collaborative Working Model to the Logistics Area using Design for Six Sigma** (Carvalho et al., 2016; *IJQRM, Emerald Group Pub. Ltd.*).
- **Investigation of Reconfiguration Effect on Makespan with Social Network Method for Flexible Job Shop Scheduling Problem** (Reddy et al., 2017, *CAIE, Elsevier*).

- **A Cloud-based Architecture with Embedded Pragmatics Renderer for Ubiquitous and Cloud Manufacturing** (Ferreira et al., 2017, *IJCIM, Taylor & Francis*)
- **Telefacturing Approach for Optimal Manufacturing Service to Enhance the Interoperability in Distributed Manufacturing Environments** (Manupati et al., 2017; *Journal of Eng., Hindawi*).
- **Web-Based Decision System for Distributed Process Planning in a Networked Manufacturing Environment** (Manupati et al., 2018; *SCI, Springer*).
- **A Novel Integrated Framework Approach for TEBC Technologies in Distributed Manufacturing Systems: A Systematic Review and Opportunities** (Ramakurthi et al., 2021a; *LNME, Springer*).
- **A Hybrid Multi-objective Evolutionary Algorithm-based Semantic Foundation for Sustainable Distributed Manufacturing Systems** (Ramakurthi et al., 2021b; *Appl. Sci., MDPI*).
- **An Innovative Approach for Resource Sharing and Scheduling in a Sustainable Distributed Manufacturing System** (Ramakurth, et al., 2022; *AEI, Elsevier*).
- **Leveraging Blockchain to Support Collaborative Distributed Manufacturing Scheduling** (Ramakurthi et al., 2023; *Sust., MDPI*).

*Intelligent and predictive managemen*t

Manufacturing managers are generally looking for models and methods that besides being able to provide efficient overall production performance further enable reactive systems to deal with unpredicted events Cardin et al., 2017; Morariu et al., 2020; Nof and Grant, 1991). According to the authors in Cardin et al. (2017), one important contribution to this end arises from the holonic/ MAS domain, which permits us to couple predictive or proactive with reactive mechanisms through agents/ holons. There are various other approaches that are being put forward for carrying out predictive or intelligent scheduling, for instance, based on different types of AI-based approaches, namely, machine/deep learning, and neural networks, among others, e.g., from the data science field, for big data processing and analysis (Morariu et al., 2020; Kalinowski et al., 2013; Jimenez et al., 2016; Sobaszek et al., 2017).

Collaborative management paradigms through the use of AI-based procedures, methods, algorithms, techniques, tools, and platforms, to ensure data forecasting or prediction, and support intelligent manufacturing management, along with security in data/information transferring, sharing and processing between entities, can be found in:

- **Smart Objects Embedded Production and Quality Management Functions** (Putnik et al., 2015; *IJQR, Univ. Montenegro*).
- **Simulation Study of Large Production Network Robustness in Uncertain Environment** (Putnik et al., 2015; *Cirp Annals–Manuf. Tech., Elsevier*).
- **A Cyber-physical System based Collaborative Distributed Manufacturing System Architecture for Intelligent Manufacturing** (Thakur et al., 2017; *Regional Helix 17*).
- **A Human Centred Hybrid MAS and Meta-Heuristics Based System for Simultaneously Supporting Scheduling and Plant Layout Adjustment** (Alves et al., 2019; *FME Transactions, Belgrade Univ.*).
- **Production Scheduling using Multi-objective Optimization and Cluster Approaches** (Azevedo et al., 2021; *LNNS, Springer*).
- **Bio-inspired Multi-objective Algorithms Applied on Production Scheduling Problems** (Azevedo et al., 2022; *IJIEC, Growing Science*).
- **Semi-double-loop Machine Learning based CPS Approach for Predictive Maintenance in Manufacturing System based on Machine Status Indications** (Putnik et al., 2021; *Cirp Annals–Manuf. Tech., Elsevier*).
- **A Self-Parametrization Framework for Meta-Heuristics** (Santos et al., 2022; *Mathematics, MDPI*).
- **Literature Review on Autonomous Production Control Methods** (Martins et al., 2020; *EIS, Taylor & Francis*).
- **Comparative Study of Autonomous Production Control Methods using Simulation** (Martins et al., 2020; *SMPT, Elsevier*).

Real-time management

Real-time-based management is of utmost importance in the I4.0, characterized by the existence of smart interconnected and communicating devices or entities (Varela et al., 2021, 2022). Thus, this paradigm is being widely used currently, and further combined with other manufacturing management paradigms and approaches (Hofer et al., 2020; Hsu and Yang, 2016; Kalinowski et al., 2013; Rahman et al., 2019). As stated in Rahman et al. (2019), with the emergence of new I4.0 technologies, real-time order acceptance and scheduling is a key problem in a make-to-order (MTO) production systems, where customers place orders in a dynamic basis, and the decision maker has promptly decide about their acceptance or rejection based on the available resources on due time. This can be achieved by combining different kinds of technology and decision- making support methods, techniques, and tools (Modekurthy et al., 2021; Chen et al., 2020; Kocsi et al., 2020; Ghaleb et al., 2020). One typical approach, among others, is based on the use of the rolling horizon technique (Hsu and Yang, 2016; Alves et al., 2021).

Collaborative management paradigm, approaches, and technology to allow real time data/information acquisition, processing, analysis, and visualization based on appropriate methods, procedures, tools, and platforms for supporting manufacturing management, in:

- **Integrated Platform for Real-time Control and Production and Productivity Monitoring and Analysis** (Vieira et al., 2016; *RRPMOM, Elsevier*).
- **Telefacturing based Distributed Manufacturing Environment for Optimal Manufacturing Service by Enhancing the Interoperability in the Hubs** (Manupati et al., 2017; *Journal of Eng., Hindawi*).
- **Intelligent Platform for Supervision and Production Activity Control in Real Time** (Vieira et al., 2018; *Adv. Manuf., Springer*).
- **Supplier Evaluation and Selection: A Fuzzy Novel Multi-Criteria Group Decision-Making Approach** (Simonov et al., 2018; *IJQR, Univ. Montenegro*).
- **Collaborative Manufacturing based on Cloud, and on Other I4.0 Oriented Principles and Technologies: A Systematic Literature Review and Reflections** (Varela et al., 2018b; *MPER, Polish Acad Sci.*).
- **How Environment Dynamics Affects Production Scheduling: Requirements for Development of CPPS Models** (Alves, Putnik, and Varela, 2021; *FME Transactions, Belgrade Univ.*).
- **Modelling and Evaluation of "Fixed Horizon", "Rolling Horizon" and "Real Time Management" Production Scheduling Paradigms in Ubiquitous Production Networks under Conditions of Dynamic Environments for Economic and Environmental Sustainability** (Alves, C., Ph.D. concluded in 2017, internal member).

Parallel management

Parallel management is not really a new paradigm (Taillard, 1994; Daniels et al., 1996; Olafsson and Shi, 2000), but another one that is gaining a refreshed attention in the current digitalization era, as it presents the potential to solve the increasingly more complex MS problems, demanding bigger computational power. In this context, it has already restarted to gain an important visibility, to continue being further explored, as parallel architectures enable high-level performance, namely, through cloud and edge computing, along with varying types of other management approaches, for instance, based on metaheuristics, along with other techniques for supporting manufacturing scheduling (Coelho and Silva, 2021; Rohaninejad et al., 2021).

Collaborative management paradigm based on the parallelization of models, methods, and procedures, by using appropriate algorithms, systems and platforms to permit parallel algorithms execution, and manufacturing management, in:

The Impact of Technological Implementation Decisions on Job-Shop Scheduling Simulator Performance using Secondary Storage and Parallel Processing (Lopes et al., 2022; *LNNS, Springer*).

Other ongoing work with a research team from the Department of Computer Science of the Polytechnic Institute of Cávado and Ave, Barcelos.

Collaborative Manufacturing Management Definition

Collaborative Manufacturing Management (CMM) can be defined as a management function that consists of planning, directing, and controlling all production factors and resources, processes, and other elements inherent in production systems and supply chains or networks, aiming at the inclusion and satisfaction of demand, in the context of a traditional or extended, distributed, virtual or networked manufacturing environments, aiming to satisfy other different conditions, to satisfy internal and external requirements and customers, based on defined business strategies, integrating all stakeholders involved, including suppliers, business partners, internal and external businesses, and clients, in order to achieve business objectives, along with sustainability and collaboration goals.

3. Collaborative Management Paradigms Applications

In this study, approximately 149 publications have been analysed, which corresponded to the set of the most relevant ones, found in the literature, satisfying a rigorous set of conditions imposed in the conducted literature search process, regarding the definition of exclusion and inclusion criteria for conducting an SLR process. Thus, this main publication set was subject to deep analysis and further discussion.

The set of the most relevant publications was reached by using the b-on platform at UMinho (https://www.b-on.pt/entidade/universidade-do-minho/). This kind of platform was chosen as it permits access to the full content of a widened collection of scientific works published in high-quality sources, for instance in journals, and in the proceeding books of international conferences, indexed in relevant scientific databases, such as the Web of Science, Scopus, Science Direct, and IEEE.

The search process was carried out by using groups of keywords about collaborative and global management in the Industry 4.0. As a result, Figure 2 shows the total amount of work obtained regarding the underlying focused management paradigms between 2011 and 2021. As can be seen in Figure 2, the CollManag paradigm that occurred the most was the integrated one, followed by the real-time one. Thereafter, the distributed paradigm appears to be highly relevant, followed narrowly by the intelligent or predictive paradigm, and next by the dynamic paradigm. A less expressive relevance does appear with the parallel paradigm, although it is a revealing and increasing tendency, as shown in Figure 3.

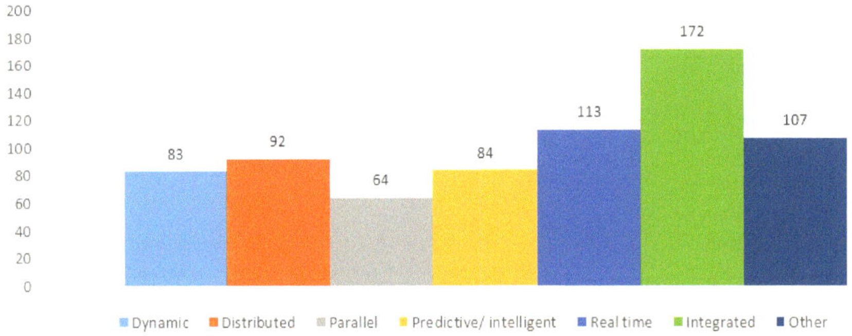

Figure 2. Number of publications about management paradigms application from 2011 to 2021 (adapted from: Varela et al., 2023).

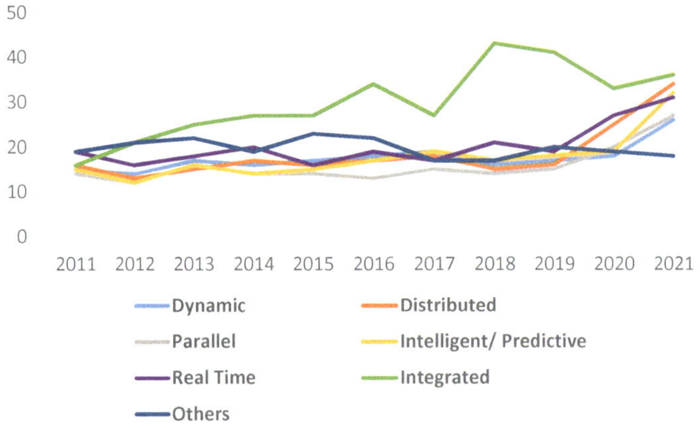

Figure 3. Evolution of the number of publications about management paradigms from 2011 to 2021 (adapted from: Varela et al., 2023).

Collaborative Management Paradigms Evolution

The evolution of the use of collaborative management paradigms over the last decade is expressed in Figure 3, regarding the progression of the CollManag about dynamic, integrated, intelligent/predictive, distributed, parallel, and real-time-based management.

It is, thus, also perceptible through Figure 3 that the integrated and the real-time paradigms are among the ones most widely being applied, and continue growing, as also happens, in general, with the remaining ones. This growing trend in the paradigms reveals its positive impact in the current digitalization era, namely, in the CollManag scope.

Moreover, the distributed and the predictive/intelligent paradigms are also receiving increased attention lately, being particularly visible, and followed by the dynamic and the parallel management ones. The last, the parallel paradigm, is the one that has been less explored during the last decade, although it is one of those that is currently experiencing a higher application increase. Furthermore, other management strategies have also been explored, namely, related with concurrent engineering applications (Demoly et al., 2013; Deshpande, 2018).

Several contributions have already been put forward, for instance, regarding the more technical aspects underlying the I4.0 context, that can promote or enhance CollManag issues in and between collaborating companies, and its inherent business and management models, namely (Sahu et al., 2018a,b,c; Bag et al., 2021; Kang et al., 2021), among others, which will be further synthetized and analysed in the next sections.

According to the study carried out, it is understandable that the CM paradigms are, to some extent, being combined. A frequent arrangement of the real-time and the dynamic, as well as with the distributed and/or with the intelligent or predictive management paradigms, is also noticeable.

Moreover, the simulation technique is also being considerably used nowadays as a CM method, and is also being combined with other technologies and approaches, namely with digital twins, and with the dynamic management paradigm, as well as other AI-based approaches, including different types of metaheuristics and MAS .

The cloud and MAS technology is also frequently being used along with metaheuristics, and with distributed, parallel, and real-time management paradigms.

Moreover, other AI-based methods, for instance, based on blockchain, smart contracts, fuzzy logic, holons, and machines of deep learning approaches, are also being frequently used, namely, in association with distributed, parallel, predictive, and real-time management paradigms, along with other approaches for enabling big data processing and analysis from the data science domain, for instance, regarding the application of intelligent and predictive management paradigms.

The rolling horizon is another frequently used approach, which is further being analysed in the scope of real-time management. Additionally, there are other kinds of methods that are being explored in the actual digitalization age, for instance, to permit other kinds of cooperative management approaches, namely, regarding manufacturing scheduling through stakeholders to reach improved solutions. Some well-known examples include the use of group decision-making models, as well as game theory, chaos and complexity analysis, and other negotiation-based management methodologies. Such kinds of approaches are frequently used in dynamic, distributed, and agile or virtual systems, or in EME (Putnik et al., 2021a,b; Eijnatten and Putnik, 2004; Arrais-Castro et al., 2018; Manupati et al., 2017; Reddy et al., 2017; Nouiri et al., 2019).

The deep analysis of the publications summarized in Table 2 further enables us to recognize that around three out of the whole set of six management paradigms are being combined. Therefore, an additional effort will be needed to properly tackle collaborative management among companies and remaining stakeholders, to reach improved decision-making processes and practices by increasing the combination of global management paradigms and underlying problem-solving approaches.

The six management paradigms underlying the proposed collaborative management meta-model previously presented are further discussed, along with other manufacturing management approaches and tools based on the bibliography analysed, in order to provide further insights and directions regarding collaborative manufacturing management research and practice in companies in the digital age.

The dynamic management paradigm is mainly characterized by the possibility of quickly adapting to changing manufacturing management conditions by adapting the corresponding management approaches and solving methods (Alves et al., 2021; Vafaei et al., 2019; Varela et al., 2018a,b; Tan et al., 2019; Ferreira et al., 2019; Saboor et al., 2019).

The distributed management paradigm is well-suited for permitting the decomposition of management problems, which may arise in the scope of extended, agile, and virtual production systems, usually characterized by higher levels of complexity associated with its underlying networked organization (Vieira et al., 2012; Ramakurthi et al., 2021; Saeidlou et al., 2019; Zhang et al., 2019; Mishra et al., 2016; Sousa and Oliveira, 2020; Ramakurthi et al., 2022).

The intelligent and predictive management paradigm is currently a main issue in the I4.0 context, and underlying CPPS, supported by AI-based approaches, plays an important role in promoting the resolution of management problems through the use of various methods and techniques that further enable us to predict data and manufacturing conditions, by exploring high volumes of varying kind of dynamically emerging data (Guo et al., 2015; Azevedo et al., 2021; Azevedo et al., 2022; Cardin et al., 2017; Sobaszek et al., 2017).

The parallel management paradigm is particularly well suited for solving 'heavy' or complex management problems through a decentralized solving methodology in which two or more entities collaborate in its resolution. The use of HPC is nowadays recommended in the I4.0 context, mainly when in the presence of big data and by further making use of compound management methods, which is quite typical in the resolution of manufacturing scheduling problems, particularly those occurring in distributed and extended manufacturing environments and which may further include CPS (Mao et al., 2020; Lopes et al., 2022).

The integrated management paradigm allows the integration of two or more management functions, for instance, regarding process planning and scheduling, batching and scheduling, scheduling and manufacturing layout arranging, scheduling and maintenance management, and scheduling and supply chain management, among other combinations, to mention just a few of the most frequently used ones (Low et al., 2013; Fu et al., 2019; Varela et al., 2021; Frazzon et al., 2018; Laili et al., 2020).

The real-time management paradigm is also one of the most popular ones in the I4.0 context, as it enables businesses to acquire, process, and analyze data in a dynamic and agile way from the manufacturing environment up to the management level through the use of appropriate technological support, based on suitable middleware, including smart objects and associated devices (Alves et al., 2021; Wang et al., 2008).

The proposed collaborative management meta-model, integrating the six main collaborative management paradigms identified, consists of original input, as the co-authors did not come across any more or less closely related work mentioning the combined use of this whole set of management paradigms either in academia or industry, as previously shown through the compiled information in Figures 2 and 3. Therefore, regarding the whole and diversified set of benefits expected through its use, further developments regarding the combination of the underlying six management paradigms is highly recommended, as each one enables us to tackle specific main issues in the context of CollManag, being considered to be of highest relevance in the current Industry 4.0 era.

Final Remark

For a full implementation of collaborative management, **the set of six main management paradigms underlying the proposed collaborative management meta-model should be jointly explored**, in order **to achieve more effective and efficient collaboration**. In this way, it will be possible to satisfy not just the inferior level of collaboration, through the use of means and technologies that permit connection, communication and sharing of some tangible or intangible asset (lower level of collaboration), but further processes and practices that allow co-learning and co-creation, in a given industrial context (higher level of collaboration).

By enabling:

Dynamic:

- Data acquisition, and pre- core and post- processing
- Maintenance tasks ranking and selection
- Manufacturing planning, programming, and control

Integrated:

- Data pre-core, and post-processing, and analysis
- Management functions (e.g., about maintenance tasks and production orders' processing and scheduling)

Distributed:

- Group decision-making
- Maintenance task ranking and selection
- Manufacturing planning, programming, and control

Intelligent/predictive:

- Data processing (about manufacturing, management, namely about maintenance management and production scheduling data, among others)
- Problems solving algorithms (e.g., production and maintenance planning and tasks scheduling, along with production control)

Parallel:

- Algorithms execution
- Decision-making

Real time:

- Data acquisition, processing, and analysis
- Management approaches

Thus, it is important to **combine, as much as possible the different paradigms** (ideally all), to achieve a higher level of:

- Collaborative engineering and management
- Sustainable development of companies
- Transition of companies to the I4.0, properly supported by underlying recent technology

Based on:

Multidisciplinary teams focused on internal and externally oriented businesses and management strategies and approaches through extended communication networks, protocols, and means to connect all stakeholders:

- ➢ Suppliers
- ➢ Internal businesses
- ➢ External business partners
- ➢ Clients

4. Contextualization of the Proposed Collaborative Management Meta-model within the State of the Art and Further Directions

According to the results obtained through the SLR conducted, it is possible to understand that the CollMang paradigms are usually combined with each other; it is visible the combination of two or more paradigms, namely the real-time with the dynamic, the distributed or the predictive ones.

According to the results obtained, it is possible to understand that the management paradigms are usually combined with each other; it is visible the combination of two or more paradigms, namely, the real-time with the dynamic, the distributed, or the predictive ones.

Simulation is currently a frequently used management approach, namely, as a scheduling one, along with digital twin, for instance, through the application of dynamic scheduling, along with a widened range of other approaches, such as from the AI domain, namely, based on diverse meta heuristics and MAS.

Cloud, MAS, along with meta heuristics, and hybrid or combined approaches are frequently appearing associated with the Distributed, the Parallel, and the Real-time-based management paradigms.

Also, a widened and diversified range of other AI-based approaches, namely, through smart contracts, fuzzy logic, holonic, and learning (machine/deep learning) based approaches are increasingly being used currently, for instance, associated to the Distributed, Parallel, Predictive, and Real-time- based management paradigms, besides other data science approaches for big data processing and analysis, namely, in Intelligent/ Predictive management scope.

The rolling-horizon is an approach that is frequently analysed in the context of real-time management. Besides, varied other approaches exist, in the current digitalization era, for enabling cooperative of collaborative management, among a more or less extended set of stakeholders for reaching joint orders schedule, for instance, based on group decision-making approaches, game theory, chaos and complexity management or negotiation, which typically occur in the context of dynamic, distributed, decentralized, virtual, extended, and agile manufacturing environments (Varela et al., 2021, 2022; Sousa and Oliveira, 2020; Delaram and Valilai, 2018; Fernandez-Viagas and Framinan, 2021; Hsu, Wang, and Chu, 2018; Nouiri, Trentesaux, and Bekrar, 2019; Tighazoui, Sauvey, and Sauer, 2021; Wenzelburger and Allgöwer, 2021; Yang and Takakuwa, 2017).

In this work the main focus consisted on studying the state of the art research about CollManag, and its relation with the collaboration theory and practice, in the current I4.0, along with the analysis of expected benefits that can arise from the combined application of management paradigms, along with different types of solving approaches, methods, and algorithms, varying from more or less pure mathematical or optimization methods up to diverse kinds of methods, such as those based on AI, for solving management problems in different production environments. These manufacturing environments can vary from more classical ones up to more recent cyber-physical and/or extended, complex, and agile or virtual manufacturing environments.

To this end, some relevant and more or less recent CollManag paradigms and underlying approaches and systems from the literature were briefly referred to, in order to better contextualize the work carried out in the scope of the I4.0 context and associated collaborative processes and practices, which aimed at a novel contribution, as no similar work was identified through the literature analysis performed.

In the I4.0 context, one typical example of CollManag is DMS, which is characterized by a set of tasks that have to be chained in order to obtain a coordinated workflow among the dispersed manufacturing resources. This chaining process results in a more or less complex production program through the allocation and sequencing of the tasks on the corresponding production resources, which has to satisfy a set of constraints related either to the production resources itself and/or to the tasks, in order to reach some simple or combined or complex goal.

Currently, due to globalization, DMS plays a crucial role, and diverse approaches have been proposed to accomplish it; a very popular one is based on a MAS, through the use of appropriate architectures and protocols (Shen, 2012).

One such contribution concerning DMS is mentioned in (Varela and Ribeiro, 2014), which is considered to be necessary in the current global production environments. Another example is presented in Varela and Ribeiro (2014) about an approach for dynamic DMS, supported by a dynamic multi-criteria decision model (DMCDM), and by further integrating strategies that enable trade-offs between diverse performance measures. Moreover, there are many various approaches, algorithms, tools, or systems and platforms to support CollManag or, such as, global manufacturing scheduling, that can be further implemented. These vary from purely centralized up to fully decentralized architectures, for instance, for further integrating other management functions, besides manufacturing scheduling, such as process planning, batching, system balancing, and layout definitions, namely referred to in the following sources (Vieira et al., 2012; Guo et al., 2015; Varela et al., 2012; Ramakurthi et al., 2021).

In Chiu and Yih (1995), a simulation model is proposed that implements a dynamic scheduling scheme to generate training scheduling examples, considered by the authors to be good schedules. Their search training was performed by using a proposed genetic algorithm, along with a tolerance-based learning algorithm requiring the acquisition of general scheduling rules from the scheduling training examples, and further adapting to new perceived examples, enabling knowledge modification. According to the authors, their experimental results showed that the dynamic scheme meaningfully outperformed a static one when integrating a simple dispatching rule for performing the distributed scheduling.

In Zhou et al. (2008), an agent-based approach is proposed for distributed manufacturing programming, which enables companies to solve a global combinatorial optimization schedule, by integrating a jobs process plan in a distributed production environment. Their approach was adapted from a particle swarm optimization (PSO) algorithm, through which the agents move towards a schedule to find the best global makespan.

Saeidlou et al. in 2019 proposed a cooperative system to perform distributed manufacturing scheduling, based on a set of rules considered to be most relevant, which are integrated through their proposed cooperative system, through an agent-based decision support system that, according to the authors, enables them to find near-optimal solutions within a reasonable computational time.

Zhang et al. in 2019 put forward an optimization algorithm centered on a discrete fruit fly optimization algorithm (DFOA), integrating an evolutionary optimization model for cost minimization, namely, energy consumption, for scheduling jobs in a distributed manufacturing system that comprises multiple factories, each one integrating a flow shop with blocking constraints. According to the authors, their proposed approach outperforms some well-known precision and convergence algorithms.

Wang, Ghenniwa, and Shen in 2008 present a real-time distributed shop floor scheduling approach, based on an agent-based service-oriented architecture, through which the shop floor is modelled as a group of flexible manufacturing systems in the form of multiple work cells. In this proposal, the authors perform the distributed scheduling process through a local dynamic scheduling approach, by the interaction of a scheduling agent, a real-time control agent, and resource agents, based on web services, for a proper integration.

Mishra et al. in 2016 describe a cloud-based multi-agent architecture for distributed manufacturing units' operational planning and scheduling. Their proposed system is self-reactive, integrated, dynamic, and autonomous, in order to assist the manufacturing industry in establishing real-time information sharing among autonomous agents, clients, suppliers, and the manufacturing units, which is illustrated through a case study.

In Fu, Wang, and Huang (2019), an integrated brainstorm optimization algorithm is put forward by the authors for distributed production, through the use of a stochastic multi-objective model. The distributed manufacturing environment consists of a set of independent flow shops with different quantities of machines. They conclude that their proposed approach can achieve satisfactory performance when compared with two other multi-objective algorithms from the literature, based on the experimental results obtained.

Mao, Li, Guo, and Wu (2020) researched cooperative planning and symmetric scheduling on parallel shipbuilding projects in the context of an open distributed manufacturing environment. To this end, the authors propose an assistant decision-making approach to support task dispatching and multi party collaboration in order to achieve better-distributed resource utilization, further helping project managers in controlling the shipbuilding practice, based on negotiation through an iterative combination auction (ICA) method for solving integrated project planning and scheduling. The authors present a demonstrative example to show the efficacy and reasonableness of their proposed approach.

Lou et al. (2010) put forward a distributed programming method supported by multi-agents for assigning tasks to machines, for being applied through a dynamic formation of virtual job-shops to satisfy manufacturing requisites, further based on market mechanisms, as well as a distributed scheduling approach based on negotiation among participating entities.

Cheng, Bi, Tao, and Ji in 2020 propose what they call a hyper network-based manufacturing service for distributed scheduling and cooperative production in smart systems, through the use of cloud services, along with real-time data, as collaborative services. Their proposed approach is further based on graph coloring and an artificial bee colony algorithm for solving the scheduling problem. The authors state that three sets of tests were performed and discussed in terms of three scenarios of distributed cooperative manufacturing processes, through a private, public, and hybrid cloud-based model.

In the concrete context of CPS, some further interesting contributions did arise. In Kim et al. (2013), a parallel programming approach is applied for analyzing a self-driving car case study.

In 2019, Nouiri, Trentesaux, and Bekrar put forward an integrated energy efficient programming approach for production systems based on a collaboration process between cyber-physical and energy systems.

Putnik and Ferreira in 2019 proposed an Industry 4.0 meta-model, which enables businesses to integrate models and tools in cyber-physical manufacturing systems.

Tan, et al. in 2019 presented an integrated approach to model, plan, and schedule operations on a shop floor assembly system characterize d by dynamic cyber-physical cooperation, which was analysed through a smart industrial robot production case study.

Another interesting contribution is referred to in Villalonga et al. (2021) about a decision-making model for supporting dynamic programming in CPPS by using digital twin technology.

5. Conclusion

In this chapter, the main results about global resources management paradigms were synthetized and analyzed, in the scope of the I4.0 era, and a collaborative management meta-model was presented and briefly described, including six paradigms concerning integrated, dynamic, intelligent/predictive, distributed, parallel, and real-time management.

The proposed manufacturing management meta-model is aimed at supporting proper collaborative management processes and practices by encompassing, as much as possible, the underlying paradigms according to specific needs of each company and associated stakeholders, in order to reach joint and enhanced decisions once around three are jointly explored. Such aims or objectives will greatly depend on the underlying manufacturing environment, which may vary from more simple, classical, or more traditional, and centralized ones up to more complex, cyber-physical, distributed, extended, and agile or virtual enterprises. These varying kinds of manufacturing environments, for instance more complex and dynamic ones, along with their underlying management strategies, assume a primer importance nowadays, in the I4.0 era, namely in the context of cyber-physical systems, as was highlighted in this chapter. It is, thus, envisioned that the proposed full joint exploration of the set of six management paradigms identified will be of the utmost importance, namely in managing such complex and highly demanding manufacturing environments as currently exist. This expectation is based on the capability of a dynamic adaptation to changing manufacturing conditions, to the decomposition or distribution and decentralization of the resolution of complex management problems, and also on the focus on different management functions through real-time-based big data acquisition, processing, and analysis in highly demanding and uncertain manufacturing environments. This is an original work, as opposed to current studies, as it permits broader and deeper insights about currently considered fundamental decision-making paradigms and underlying approaches, which are, thus, suggested to be further explored and combined, to enable to businesses to properly support manufacturing management, to carry it out in a collaborative manner, and to further support and promote the current Industry 4.0 technology. However, some limitations are expected to occur, related to the joint exploration of different kinds of management paradigms, due to the underlying highly-demanding knowledge and technology for permitting a full exploration of its joint application, along with associated problem-solving approaches. Thus, additional future work is suggested, namely for finding out some promising technologies for enabling the proper application of the proposed collaborative management framework in real industrial and academic scenarios, through the combined use of its six management paradigms, along with suitable approaches and tools for permitting prosperous and true innovation and company development in the current digitalization era.

Further, manufacturing management can be carried out through the application of different kinds of paradigms and approaches. In this chapter, the current I4.0 dynamic, distributed or decentralized, parallel, predictive or intelligent, and real-time based ones were explored and discussed, based on results obtained through a systematic literature review. The results demonstrated not only their increasing importance nowadays, but further that those paradigms and approaches are frequently combined. The dynamic and the real-time based management paradigms are frequently combined with some others, namely with the distributed or the predictive ones.

Concerning supporting technologies, the cloud, multi-agent systems, along with meta-heuristics, and hybrid or combined approaches are also increasingly being used, appearing most frequently associated with the Distributed, the Parallel, and the Real-time based management paradigms. A widened and diversified range of other AI-based approaches are increasingly being used, for instance associated to the Distributed, Parallel, Predictive, and Real-time-based management paradigms, besides other data science approaches for big data processing and analysis, namely in Predictive or Intelligent management. The rolling-horizon is an approach that is frequently analysed in the context of real-time management.

Besides, other approaches do exist, and are assuming an increasing importance within I4.0. For instance to enable cooperative or collaborative management among a more or less extended set of companies, to jointly prepare manufacturing plans and schedules, for instance based on group decision-making or negotiation, which are of particular interest in distributed, virtual, extended and agile manufacturing environments.

References

Akbar, M., and Irohara, T. Scheduling for sustainable manufacturing: A review. Journal of Cleaner Production, 2018, *205*, 2018, 866–883. Gahm, C., Denz, F., Dirr, M., and Tuma, A. Energy-efficient scheduling in manufacturing companies: A review and research framework. European Journal of Operational Research, 2016, *248*, 744–757.

Alves, C., Putnik, G. D., and Varela, M. L. R. How environment dynamics affects production scheduling: Requirements for development of CPPS models. FME Trans., 2021, *49*, 827–834.

Alves, C., Putnik, G. D., and Varela, M. L. R. How environment dynamics affects production scheduling: Requirements for development of CPPS models. FME Transactions, 2021, *49*, 827–834.

Alves, F., Varela, M. L. R., Rocha, A. M. A., Pereira, A. I., and Leitão, P. A human- centered hybrid MAS and meta heuristics-based system for simultaneously supporting scheduling and plant layout adjustment. FME Trans., 2019, *47*, 699–710.

Arrais-Castro, A., Varela, M. L. R., Putnik, G. D., Ribeiro, R. A., Machado, J., and Ferreira, L. Collaborative framework for virtual organisation synthesis based on a dynamic multi-criteria decision model. Int. J. Comput. Integr. Manuf., 2018, *31*, 857–868.

Azevedo, B. F., Montaño-Vega, R., Varela, M. L. R., and Pereira, A. I. Bio-Inspired Multi-Objective Algorithms Applied on Production Scheduling Problems. Int. J. Ind. Eng. Comput., 2022, *6*, 145–156. https://doi.org/10.5267/j.ijiec.2022.12.001.

Azevedo, B. F., Varela, M. L. R., and Pereira, A. I. Production Scheduling Using Multi-objective Optimization and Cluster Approaches. In: Proceedings of the International Conference on Innovations in Bio-Inspired Computing and Applications, 16–18 December, 2021, On the World Wide Web; Springer: Cham, Switzerland, 2021; pp. 120–129.

Bag, S., Sahu, A. K., Kilbourn, P., Pisa, N., Dhamija, P., and Sahu, A. K. Modeling barriers of digital manufacturing in a circular economy for enhancing sustainability. Int. J. Product. Perform. Manag., 2021, *71*, 833–869.

Bolton, W. Mechatronics Electronic Control Systems in Mechanical and Electrical Engineering. Pearson Education, England. 2008.

Canadas, N., Machado, J., Soares, F., Barros, C., and Varela, L. Simulation of cyber physical systems behaviour using timed plant models. Mechatronics, 2018, *54*, 175–185. https://doi.org/10.1016/j.mechatronics.2017.10.009.

Cardin, O., Trentesaux, D., Thomas, A., Castagna, P., Berger, T., and Bril El-Haouzi, H. Coupling predictive scheduling and reactive control in manufacturing hybrid control architectures: State of the art and future challenges. J. Intell. Manuf., 2017, *28*, 1503–1517.

Cardin, O., Trentesaux, D., Thomas, A., Castagna, P., Berger, T., and Bril El-Haouzi, H. Coupling predictive scheduling and reactive control in manufacturing hybrid control architectures: State of the art and future challenges. Journal of Intelligent Manufacturing, 2017, *28*, 1503–1517. Morariu, C., Morariu, O., Raileanu, S., and Borangiu, T. Machine learning for predictive scheduling and resource allocation in large scale manufacturing systems. Computers in Industry, 2020, 120, 4.

Carvalho, M. S., Magalhaes, D. S., Varela, M. L. R., Sa, J. O., and Gonçalves, I. Definition of a collaborative working model to the logistics area using design for Six Sigma. Int. J. Qual. Reliab. Manag., 2016, *43*, 465–475.

Chen, D. Enterprise-control system integration—An international standard. Int. J. Prod. Res., 2005, *43*, 4335–4357.

Chen, S., Fang, S., and Tang, R. An ANN-based approach for real-time scheduling in cloud manufacturing. Appl. Sci., 2020, 10, 2491.

Chen, S., Fang, S., and Tang, R. An ANN-based approach for real-time scheduling in cloud manufacturing. Applied Sciences, 2020, 10, 2491.

Cheng, Y., Bi, L., Tao, F., and Ji, P. Hypernetwork-based manufacturing service scheduling for distributed and collaborative manufacturing operations towards smart manufacturing. J. Intell. Manuf., 2020, *31*, 1707–1720.

Chiu, C., and Yih, Y., A learning-based methodology for dynamic scheduling in distributed manufacturing systems. Int. J. Prod. Res., 1995, *33*, 3217–3232.

Coelho, P., and Silva, C. Parallel Metaheuristics for shop scheduling: Enabling industry 4.0. Procedia Comput. Sci., 2021, *180*, 778–786.

D'Aniello, G., De Falco, M., and Mastrandrea, N. Designing a multi-agent system architecture for managing distributed operations within cloud manufacturing. Evol. Intell., 2021, *14*, 2051–2058.

Daniels, R. L., Hoopes, B. J., and Mazzola, J. B. Scheduling parallel manufacturing cells with resource flexibility. Management Science, 1996, *42*, 1260–1276.

Delaram, J., and Valilai, O. F. A mathematical model for task scheduling in cloud manufacturing systems focusing on global logistics. Procedia Manuf., 2018, *17*, 387–394.

Delaram, J., and Valilai, O. F. A mathematical model for task scheduling in cloud manufacturing systems focusing on global logistics. Procedia Manufacturing, 2018, 17, 387–394.

Deloitte. Global Powers of Retailing 2014. Retail beyond Begins. 2014. Available online: http://www.deloitte.com/(accessed on 12 September 2022).

Deloitte. Retail Globalization. Deloitte Touche Tohmatsu Limited. 2012. Available online: https://www.deloitte.com/global/en.html (accessed on 17 October 2022).

Demoly, F., Dutartre, O., Yan, X. T., Eynard, B., Kiritsis, D., and Gomes, S. Product relationships management enabler for concurrent engineering and product lifecycle management. Comput. Ind., 2013, *64*, 833–848.

Deshpande, A. Concurrent engineering, knowledge management, and product innovation. J. Oper. Strateg. Plan., 2018, *1*, 204–231.

Dogan, A., and Birant, D. Machine learning and data mining in manufacturing. Expert Syst. Appl., 2021, 166, 114060.

Dogan, A., and Birant, D. Machine learning and data mining in manufacturing. Expert Systems with Applications, 2021, 166.

Ebufegha, A., and Li, S. Multi-agent system model for dynamic scheduling in flexib le job shops. In: Proceedings of the 2021 Winter Simulation Conference (WSC), Phoenix, AZ, USA, 12–15 December 2021; pp. 1–12.

Ebufegha, A., and Li, S. Multi-agent system model for dynamic scheduling in flexib le job shops. In: 2021 Winter Simulation Conference (WSC); 2021, pp. 1–12.

Eijnatten, F. M., and Putnik, G. D. Chaos, complexity, learning, and the learning organization: Towards a chaordic enterprise. Learn. Organ., 2004, *11*, 418–429.

Fernandez-Viagas, V., and Framinan, J. M. Exploring the benefits of scheduling with advanced and real-time information integration in Industry 4.0: A computational study. J. Ind. Inf. Integr., 2021, 27, 100281.

Fernandez-Viagas, V., and Framinan, J. M. Exploring the benefits of scheduling with advanced and real-time information integration in Industry 4.0: A computational study. Journal of Industrial Information Integration, 2021, 100281.

Ferreirinha, L., Baptista, S., Pereira, Â., Santos, A. S., Bastos, J., Madureira, A. M., and Varela, M.L.R. An Industry 4.0 oriented tool for supporting dynamic selection of dispatching rules based on Kano model satisfaction scheduling. FME Transactions, 2019, 47, 757–764.

Ferreirinha, L., Baptista, S., Pereira, Â., Santos, A. S., Bastos, J., Madureira, A. M., and Varela, M. L. R. An Industry 4.0 oriented tool for supporting dynamic selection of dispatching rules based on Kano model satisfaction scheduling. FME Trans., 2019, 47, 757–764.

Framinan, J. M., and Ruiz, R. Architecture of manufacturing scheduling systems: Literature review and an integrated proposal. European Journal of Operational Research, 2010, 205, 237–246. https://doi.org/10.1016/j.ejor.2009.09.026.

Frazzon, E. M., Albrecht, A., Pires, M., Israel, E., Kück, M., and Freitag, M. Hybrid approach for the integrated scheduling of production and transport processes along supply chains. Int. J. Prod. Res., 2018, *56*, 2019–2035.

Fu, Y., Wang, H., and Huang, M. Integrated scheduling for a distributed manufacturing system: A stochastic multi-objective model. Enterp. Inf. Syst., 2019, *13*, 557–573.

Gahm, C., Denz, F., Dirr, M., and Tuma, A. Energy-efficient scheduling in manufacturing companies: A review and research framework. Eur. J. Oper. Res., 2016, 248, 744–757.

Ghaleb, M., Zolfagharinia, H., and Taghipour, S. Real-time production scheduling in the Industry-4.0 context: Addressing uncertainties in job arrivals and machine breakdowns. Comput. Oper. Res., 2020, 123, 105031.

Ghaleb, M., Zolfagharinia, H., and Taghipour, S. Real-time production scheduling in the Industry-4.0 context: Addressing uncertainties in job arrivals and machine breakdowns. Computers & Operations Research, 2020, 123, 1.

Giret, A., Trentesaux, D., and Prabhu, V. Sustainability in manufacturing operations scheduling: A state of the art review. Journal of Manufacturing Systems, 37, 2015, 126–140. https://doi.org/https://doi.org/10.1016/j.jmsy.2015.08.002.

Guo, Z., Ngai, E., Yang, C., and Liang, X. An RFID-based intelligent decision support system architecture for production monitoring and scheduling in a distributed manufacturing environment. Int. J. Prod. Econ., 2015, 159, 16–28.

Hankel, M., and Rexroth, B. The reference architectural model industrie 4.0 (rami 4.0). ZVEI, 2015, 2, 4–9.

Hofer, F., Sehr, M. A., Russo, B., and Sangiovanni-Vincentelli, A. ODRE Workshop: Probabilistic Dynamic Hard Real-Time Scheduling in HPC. In: IEEE 23rd International Symposium on Real-Time Distributed Computing (ISORC); 2020, pp. 207–212. IEEE.

Hofer, F., Sehr, M. A., Russo, B., and Sangiovanni-Vincentelli, A. ODRE Workshop: Probabilistic Dynamic Hard Real-Time Scheduling in HPC. In: Proceedings of the IEEE 23rd International Symposium on Real-Time Distributed Computing (ISORC), Nashville, TN, USA 19–21 May 2020; IEEE: Piscataway, NJ, USA, 2020; pp. 207–212.

Hsu, C. H., and Yang, H. C. Real-time near-optimal scheduling with rolling horizon for automatic manufacturing cell. IEEE Access, 2016, *5*, 3369–3375.

Hsu, C. H., and Yang, H. C. Real-time near-optimal scheduling with rolling horizon for automatic manufacturing cell. IEEE Access, 2016, *5*, 3369–3375. Zhang, J., Ding, G., Zou, Y., Qin, S., and Fu, J. Review of job shop scheduling research and its new perspectives under Industry 4.0. Journal of Intelligent Manufacturing, *30*, 2019, 1809–1830.

Hsu, T. H., Wang, L. C., and Chu, P. C. Development of a cloud-based advanced planning and scheduling system. Procedia Manuf., 2018, *17*, 427–434.

Hsu, T. H., Wang, L. C., and Chu, P. C. Development of a cloud-based advanced planning and scheduling system. Procedia Manufacturing, 2018, *17*, 427–434.

Jiang, Z., Yuan, S., Ma, J., and Wang, Q. The evolution of production scheduling from Industry 3.0 through Industry 4.0. International Journal of Production Research, 2021, 1–21. D'Aniello, G., De Falco, M., and Mastrandrea, N. Designing a multi-agent system architecture for managing distributed operations within cloud manufacturing, 2021, *14*, 2051–2058.

Jimenez, J. F., Bekrar, A., Trentesaux, D., and Leitão, P. A switching mechanism framework for optimal coupling of predictive scheduling and reactive control in manufacturing hybrid control architectures. Int. J. Prod. Res., 2016, *54*, 7027–7042.

Jimenez, J. F., Bekrar, A., Trentesaux, D., and Leitão, P. A switching mechanism framework for optimal coupling of predictive scheduling and reactive control in manufacturing hybrid control architectures. International Journal of Production Research, 2016, 54, 7027–7042.

Kagermann, H. Change through digitization—Value creation in the age of Industry 4.0. In: Management of Permanent Change; Springer Gabler: Wiesbaden, Germany, 2015; pp. 23–45.

Kalinowski, K., and Krenczyk, D., and Grabowski, C. Predictive-reactive strategy for real-time scheduling of manufacturing systems. Applied Mechanics and Materials, 2013, 307, 470–473.

Kalinowski, K., Krenczyk, D., and Grabowski, C. Predictive-reactive strategy for real time scheduling of manufacturing systems. Appl. Mech. Materials, 2013, *307*, 470–473.

Kang, D., Prabhu, M., Ahmed, R. R., Zhang, Z., and Sahu, A. K. Digital-IIoTs spheres approach toward public development: An exploiting fuzzy-grey mathematical modeling of IIoTs spheres. Grey Syst. Theory Appl., 2021, *12*, 389–416.

Kays, H. M. E., Karim, A. N. M., Daud, M. R. C., Varela, M. L. R., Putnik, G. D., and Machado, J. M. A collaborative multiplicative Holt-Winters forecasting approach with dynamic fuzzy-level components. Appl. Sci., 2018, 8, 530. https://doi.org/10.3390/app8040530.

Kim, J., Kim, H., Lakshmanan, K., and Rajkumar, R. Parallel scheduling for cyber-physical systems: Analysis and case study on a self-driving car. In: Proceedings of the ACM/IEEE 4th International Conference on Cyber-Physical Systems, Philadelphia, PA, USA, 8–11 April 2013; pp. 31–40.

Kitchenham, B. Procedures for Performing Systematic Reviews, Version 1.0. Empirical Software Engineering, 2004, *33*, 1–26.

Kitchenham, B., Brereton, O. P., Budgen, D., Turner, M., Bailey, J., and Linkman, S. Systematic literature reviews in software engineering—a systematic literature review. Information and Software Technology, 2009, *51*, 7–15.

Kocsi, B., Matonya, M. M., Pusztai, L. P., and Budai, I. Real-time decision-support system for high-mix low-volume production scheduling in industry 4.0. Processes, 2020, *8*, 912.

Kocsi, B., Matonya, M. M., Pusztai, L. P., and Budai, I. Real-time decision-support system for high-mix low-volume production scheduling in industry 4.0. Processes, 2020, *8*, 912.

Laili, Y., Lin, S., and Tang, D. Multi-phase integrated scheduling of hybrid tasks in cloud manufacturing environments. Robot. Comput.-Integr. Manuf., 2020, 61, 101850.

Leitão P., and R. F. A holonic approach to dynamic manufacturing scheduling. Robotics and Computer-Integrated Manufacturing, 2008, *24*, 625–634.

Leusin, M., Frazzon, E., Uriona Maldonado, M., Kück, M., and Freitag, M. Solving the Job-Shop Scheduling Problem in the Industry 4.0 Era. Technologies, 2018, *6*, 107.

Li Da Xu, E. L. X. and L. L. Industry 4.0: State of the art and future trends. International Journal of Production Research, *56*, 2018, 2941–2962.

Li, L. China's manufacturing locus in 2025: With a comparison of "Made-in-China 2025" and "Industry 4.0". Technol. Forecast. Soc. Change, 2018, *135*, 66–74.

Liao, Y., Loures, E. D. F. R., and Deschamps, F. Industrial Internet of Things: A systematic literature review and insights. IEEE Internet Things J., 2018, *5*, 4515–4525.

Lin, S.-W., Miller, B., Durand, J., Joshi, R., Didier, P., Chigani, A., Torenbeek, R., Duggal, D., Martin, R., Bleakley, G., et al. Industrial internet reference architecture. Ind. Internet Consort. (IIC) Tech. Rep. 2015, 1–19.

Liu, Y., Wang, L., Wang, X. V., Xu, X., and Zhang, L. Scheduling in cloud manufacturing: State-of-the-art and research challenges. Int. J. Prod. Res,. 2019, *57*, 4854–4879. https://doi.org/10.1080/00207543.2018.1449978.

Liu, Y., Wang, L., Wang, X. V., Xu, X., and Zhang, L. Scheduling in cloud manufacturing: state-of-the-art and research challenges. International Journal of Production Research, 2019, 57, 4854–4879. https://doi.org/10.1080/00207543.2018.1449978.

Lohmer, J., and Lasch, R. Production planning and scheduling in multi-factory production networks: A systematic literature review. Int. J. Prod. Res., 2021, *59*, 2028–2054.

Lohmer, J., and Lasch, R. Production planning and scheduling in multi-factory production networks: a systematic literature review. International Journal of Production Research, 2021, 59, 2028–2054. Taillard, E. D. Parallel taboo search techniques for the job shop scheduling problem. ORSA Journal on Computing, 1994, *6*, 108–117.

Lopes, N., Costa, B., Alves, C. F., Putnik, G. D., Varela, M. L. R., Cruz-Cunha, M. M., and Ferreira, L. The Impact of Technological Implementation Decisions on Job-Shop Scheduling Simulator Performance using Secondary Storage and Parallel Processing. 1st International Symposium on Industrial Engineering and Automation (ISIEA 2022), Managing and Implementing the Digital Transformation, 21st–22nd June 2022, Bozen-Bolzano, Italy. Lecture Notes in Networks and Systems, Springer 2022, 227–236.

Lopes, N., Costa, B., Alves, C. F., Putnik, G. D., Varela, L., Cruz-Cunha, M. M., and Ferreira, L. The Impact of Technological Implementation Decisions on Job-Shop Scheduling Simulator Performance Using Secondary Storage and Parallel Processing. In: International Symposium on Industrial Engineering and Automation; Springer: Cham, Switzerland, 2022; pp. 227–236.

Lou, P., Ong, S. K., and Nee, A. Y. C. Agent-based distributed scheduling for virtual job shops. Int. J. Prod. Res., 2010, *48*, 3889–3910.

Low, C., Li, R. K., and Chang, C.M. Integrated scheduling of production and delivery with time windows. Int. J. Prod. Res., 2013, *51*, 897–909.

Manupati, V. K., Gokula Krishnan, M., Varela, M. L. R., and Machado, J. Telefacturing based distributed manufacturing environment for optimal manufacturing service by enhancing the interoperability in the hubs. J. Eng., 2017, 2017, 9305989. https://doi.org/10.1155/2017/9305989.

Mao, X., Li, J., Guo, H., and Wu, X. Research on Collaborative Planning and Symmetric Scheduling for Parallel Shipbuilding Projects in the Open Distributed Manufacturing Environment. Symmetry, 2020, 12, 161.

Mishra, N., Singh, A., Kumari, S., Govindan, K., and Ali, S. I. Cloud-based multi-agent architecture for effective planning and scheduling of distributed manufacturing. Int. J. Prod. Res., 2016, *54*, 7115–7128.

Modekurthy, V. P., Saifullah, A., and Madria, S. A Distributed Real-time Scheduling System for Industrial Wireless Networks. ACM Trans. Embed. Comput. Syst. (TECS), 2021, *20*, 1–28.

Modekurthy, V. P., Saifullah, A., and Madria, S. A Distributed Real-time Scheduling System for Industrial Wireless Networks. ACM Transactions on Embedded Computing Systems (TECS), 2021, *20*, 1–28.

Monostori, L. Cyber-physical production systems: Roots, expectations, and R&D challenges. Procedia CIRP, 2014, *17*, 9–13. https://doi.org/10.1016/j.procir.2014.03.115.

Moon, J.-Y., and Park, J. Smart production scheduling with time-dependent and machine-dependent electricity cost by considering distributed energy resources and energy storage. Int. J. Prod. Res., 2014, *52*, 3922–3939. https://doi.org/10.1080/00207543.2013.860251.

Morariu, C., Morariu, O., Raileanu, S., and Borangiu, T. Machine learning for predictive scheduling and resource allocation in large scale manufacturing systems. Comput. Ind., 2020, *120*, 103244.

Nof, S. Y., and Hank Grant, F. Adaptive/predictive scheduling: Review and a general framework. Production Planning and Control, 1991, *2*, 298–312.

Nouiri, M., Trentesaux, D., and Bekrar, A. Towards energy efficient scheduling of manufacturing systems through collaboration between cyber physical production and energy systems. Energies, 2019, *12*, 4448.

Nouiri, M., Trentesaux, D., and Bekrar, A. Towards energy efficient scheduling of manufacturing systems through collaboration between cyber physical production and energy systems. Energies, 2019, 12, 4448.

Olafsson, S., and Shi, L. A method for scheduling in parallel manufacturing systems with flexible resources. IIE Transactions, 2000, 32, 135–146. Coelho, P., and Silva, C. Parallel Metaheuristics for shop scheduling: Enabling industry 4.0. Procedia Computer Science, 2021, *180*, 778–786.

Prades, L., Romero, F., Estruch, A., García-Domínguez, A., and Serrano, J. Defining a methodology to design and implement business process models in BPMN according to the standard ANSI/ISA-95 in a manufacturing enterprise. Procedia Eng., 2013, *63*, 115–122.

Putnik, G. D., and Cruz-Cunha, M. M. Knowledge and Technology Management in Virtual Organizations: Issues, Trends, Opportunities, and Solutions: Issues, Trends, Opportunities and Solutions; IGI Global: Hershey, PA, USA, 2006.

Putnik, G. D., and Cruz-Cunha, M. M. Virtual Enterprise Integration: Technological and Organizational Perspectives: Technological and Organizational Perspectives; IGI Global: Hershey, PA, USA, 2005.

Putnik, G. D., and Ferreira, L. G. M. Industry 4.0: Models, tools and cyber-physical systems for manufacturing. FME Trans., 2019, 47, 659–662.

Putnik, G. D., and Varela, M. L. R. How environment dynamics affects production scheduling: Requirements for development of CPPS models. FME Trans., 2021, 49, 827–834.

Putnik, G. D., Putnik, Z., Shah, V., Varela, L., Ferreira, L., Castro, H., Catia, A., and Pinheiro, P. Collaborative Engineering definition: Distinguishing it from Concurrent Engineering through the complexity and semiotics lenses. In: IOP Conference Series: Materials

Science and Engineering; IOP Publishing: Bristol, UK, 2021; Volume 1174, No. 1, p. 012027.

Putnik, G. D., Putnik, Z., Shah, V., Varela, L., Ferreira, L., Castro, H., Catia, A., and Pinheiro, P. Collaborative Engineering: A Review of Organisational Forms for Implementation and Operation. In: IOP Conference Series: Materials Science and Engineering; IOP Publishing: Bristol, UK, 2021; Volume 1174, No. 1, p. 012028.

Putnik. G. D., Pabba, S. K., Manupati, V. K., Varela, M. L. R., and Ferreira, F. Semi-Double-loop machine learning based CPS approach for predictive maintenance in manufacturing systems based on machine status indications. CIRP Ann.-Manuf. Technol., 2021, *70*, 365–368. https://doi.org/10.1016/j.cirp.2021.04.046.

Rahman, H. F., Janardhanan, M. N., and Nielsen, I. E. Real-time order acceptance and scheduling problems in a flow shop environment using hybrid GA-PSO algorithm. IEEE Access, 2019, *7*, 11275–12742.

Rahman, H. F., Janardhanan, M. N., and Nielsen, I. E. Real-time order acceptance and scheduling problems in a flow shop environment using hybrid GA-PSO algorithm. IEEE Access, 2019, *7*, 11275–12742.

Ramakurthi, V., Manupati, V. K., Machado, J., Varela, L., and Babu, S. An innovative approach for resource sharing and scheduling in a sustainable distributed manufacturing system. Adv. Eng. Inform., 2022, *52*, 101620.

Ramakurthi, V. B., Manupati, V. K., Machado, J., and Varela, L. A hybrid multi-objective evolutionary algorithm-based semantic foundation for sustainable distributed manufacturing systems. Appl. Sci., 2021, *11*, 6314.

Reddy, M. S., Ratnam, C., Agrawal, R., Varela, M. L. R., Sharma, I., and Manupati, V.K. Investigation of reconfiguration effect on makespan with social network method for flexible job shop scheduling problem. Comput. Ind. Eng., 2017, *110*, 231–241.

Rohaninejad, M., Tavakkoli-Moghaddam, R., Vahedi-Nouri, B., Hanzalek, Z., and Shirazian, S. A hybrid learning-based meta heuristic algorithm for scheduling of an additive manufacturing system consisting of parallel SLM machines. Int. J. Prod. Res., 2021, *60*, 6205–6225.

Rohaninejad, M., Tavakkoli-Moghaddam, R., Vahedi-Nouri, B., Hanzalek, Z., and Shirazian, S. A hybrid learning-based meta-heuristic algorithm for scheduling of an additive manufacturing system consisting of parallel SLM machines. International Journal of Production Research, 2021, 1–21.

Rossit, D.A., Tohmé, F., and Frutos, M. Industry 4.0: Smart Scheduling. Int. J. Prod. Res., 2019, 57, 3802–3813. https://doi.org/10.1080/00207543.2018.1504248.

Saboor, A., Imran, M., Agha, M. H. and Ahmed, W. Flexible cell formation and scheduling of robotics coordinated dynamic cellular manufacturing system: A gateway to Industry 4.0. In: Proceedings of the 2019 International Conference on Robotics and Automation in Industry, Rawalpindi, Pakistan, 21–22 October 2019; IEEE: Piscataway, NJ, USA, 2019; pp. 1–6.

Saeidlou, S., Saadat, M., Amini Sharifi, E., and Jules, G. D. Agent-based distributed manufacturing scheduling: An ontological approach. Cogent Eng., 2019, *6*, 1565630.

Sahu, A. K., Narang, H. K., Rajput, M. S., Sahu, N. K., Sahu, A. K. Performance modeling and benchmarking of green supply chain management: An integrated fuzzy approach. Benchmarking Int. J., 2018, *25*, 2248–2271.

Sahu, A. K., Sahu, A. K., Sahu, N. K. A review on the research growth of industry 4.0: IIoT business architectures benchmarking. Int. J. Bus. Anal. (IJBAN), 2020, *7*, 77–97.

Sahu, N. K., Sahu, A. K., and Sahu, A. K. Green supply chain management assessment under chains of uncertain indices: An intellectual approach. J. Model. Manag., 2018, *13*, 973–993.

Seuring, S., and Müller, M. Core issues in sustainable supply chain management: A Delphi study. Bus. Strategy Environ., 2008, *17*, 455–466.

Shen, W. Distributed manufacturing scheduling using intelligent agents. IEEE Intell. Syst,. 2002, *17*, 88–94.

Sobaszek L., G. A. and S. A. Predictive scheduling as a part of an intelligent job scheduling system. In: International Conference on Intelligent Systems in Production Engineering and Maintenance. Springer. 2017.

Sobaszek, L., Gola, A., and Świć, A. Predictive scheduling as a part of an intelligent job scheduling system. In: Proceedings of the International Conference on Intelligent Systems in Production Engineering and Maintenance, Wroclaw, Poland, 28–29 September 2017; Springer: Berlin/Heidelberg, Germany, 2017.

Soldatos, J., Gusmeroli, S., Malo, P., and Di Orio, G. Internet of things applications in future manufacturing. In: Digitising Industry-Internet of Things Connecting the Physical, Digital and Virtual Worlds; River Publishers: New York, NY, USA, 2016; pp. 153–183.

Sony, M., and Naik, S. S. Ten lessons for managers while implementing Industry 4.0. IEEE Eng. Manag. Rev., 2019, 47, 45–52.

Sousa, A. L., and Oliveira, A. S. Distributed MAS with Leaderless Consensus to Job-Shop Scheduler in a Virtual Smart Factory with Modular Conveyors. In: Proceedings of the 2020 Latin American Robotics Symposium (LARS), 2020 Brazilian Symposium on Robotics (SBR) and 2020 Workshop on Robotics in Education (WRE), Natal, Brazil, 9–13 November 2020; IEEE: Piscataway, NJ, USA, 2020; pp. 1–6.

Sousa, A. L., and Oliveira, A. S. Distributed MAS with Leaderless Consensus to Job-Shop Scheduler in a Virtual Smart Factory with Modular Conveyors. In: 2020 Latin American Robotics Symposium (LARS), 2020 Brazilian Symposium on Robotics (SBR) and 2020 Workshop on Robotics in Education (WRE); 2020, pp. 1–6. IEEE.

Tan, Q., Tong, Y., Wu, S., and Li, D. Modeling, planning, and scheduling of shop-floor assembly process with dynamic cyber-physical interactions: A case study for CPS-based smart industrial robot production. Int. J. Adv. Manuf. Technol., 2019, *105*, 3979–3989.

Tan, Q., Tong, Y., Wu, S., and Li, D. Modelling, planning, and scheduling of shop-floor assembly process with dynamic cyber-physical interactions: A case study for CPS-based smart industrial robot production. The International Journal of Advanced Manufacturing Technology, 2019, *105*, 3979–3989.

Tighazoui, A., Sauvey, C., and Sauer, N. Predictive-reactive Strategy for Flowshop Rescheduling Problem: Minimizing the Total Weighted Waiting Times and Instability. J. Syst. Sci. Syst. Eng., 2021, *30*, 253–275.

Tighazoui, A., Sauvey, C., and Sauer, N. Predictive-reactive Strategy for Flowshop Rescheduling Problem: Minimizing the Total Weighted Waiting Times and Instability. Journal of Systems Science and Systems Engineering, 2021, *30*, 253–275.

Tranfield, D., Denyer, D., and Smart, P. Towards a methodology for developing evidence-informed management knowledge by means of systematic review. Br. J. Manag., 2003, *14*, 207–222.

Vafaei, N., Ribeiro, R. A., Camarinha-Matos, L. M., and Varela, L. R. Normalization techniques for collaborative networks. Kybernetes, 2019, *49*(4), 1285–1304. https://doi.org/10.1108/K-09-2018-0476.

Varela, L., Ávila, P., Castro, H., Putnik, G. D., Fonseca, L. M. C., and Ferreira, L. Manufacturing and Management Paradigms, Methods and Tools for Sustainable Industry 4.0-Oriented Manufacturing Systems. Sustainability, 2022, *14*, 1574. https://doi.org/10.3390/su14031574.

Varela, M. L., Madureira, A. M., Dantas, J. D. Santos, A. S., Putnik, G. D., Trojanowska, J., and Machado, J. Collaborative paradigm for single-machine scheduling under just-in-time principles: Total holding-tardiness cost problem. Manag. Prod. Eng. Rev., 2018, *9*(1), 90–103.

Varela, M. L., Putnik, G. D., Manupati, V. K., Rajyalakshmi, G., Trojanowska, J., and Machado, J. Integrated process planning and scheduling in networked manufacturing systems for I4. 0: A review and framework proposal. Wireless Networks, 2021, *27*, 1587–1599.

Varela, M. L. R., Alves, C. F. V., Santos, A. S., Vieira, G. G., Lopes, N., and Putnik, G. D. Analysis of a Collaborative Scheduling Model Applied in a Job Shop Manufacturing Environment. Machines, 2022, *10*, 1138. https://doi.org/10.3390/machines10121138.

Varela, M. L. R., and Ribeiro, R. A. Distributed Manufacturing Scheduling Based on a Dynamic Multi-criteria Decision Model. In: Recent Developments and New Directions in Soft Computing. Studies in Fuzziness and Soft Computing; Zadeh, L., Abbasov, A., Yager, R., Shahbazova, S., and Reformat, M. (Eds.). Springer: Berlin/Heidelberg, Germany, 2014; Volume 317. https://doi.org/10.1007/978-3-319-06323-2_6.

Varela, M. L. R., Putnik, G. D., and Romero, F. The Concept of Collaborative Engineering: A Systematic Literature Review. Prod. Manuf. Res., 2022, *10*, 784–839. https://doi.org/10.1080/21693277.2022.2133856.

Varela, M. L. R., Putnik, G. D., Manupati, V. K., Rajyalakshmi, G., Trojanowska, J., and Machado, J. Collaborative manufacturing based on cloud, and on other I4.0 oriented principles and technologies: A systematic literature review and reflections. Manag. Prod. Eng. Rev., 2018, *9*, 90–99. https://doi.org/10.24425/119538.

Varela, M. L. R., Putnik, G. D., Manupati, V. K., Rajyalakshmi, G., Trojanowska, J., and Machado, J. Integrated process planning and scheduling in networked manufacturing systems for I4.0: A review and framework proposal. Wirel. Netw., 2021, *27*, 1587–1599. https://doi.org/10.1007/s11276-019-02082-8.

Varela, M. L. R., Putnik, G. D., and Cruz-Cunha, M. M. Web-based technologies integration for distributed manufacturing scheduling in a virtual enterprise. Int. J. Web Portals (IJWP), 2012, *4*, 19–34.

Vieira, G., Varela, M. L. R., Putnik, G. D. Technologies integration for distributed manufacturing scheduling in a virtual enterprise. In: Proceedings of the International Conference on Virtual and Networked Organizations, Emergent Technologies, and Tools, Ofir, Portugal, 6–8 July 2012; Springer: Berlin/Heidelberg, Germany, 2012; pp. 337–347.

Villalonga, A., Negri, E., Biscardo, G., Castano, F., Haber, R. E., Fumagalli, L., and Macchi, M. A decision-making framework for dynamic scheduling of cyber-physical production systems based on digital twins. Annu. Rev. Control, 2021, *51*, 357–373.

Villalonga, A., Negri, E., Biscardo, G., Castano, F., Haber, R. E., Fumagalli, L., and Macchi, M. A decision-making framework for dynamic scheduling of cyber-physical production systems based on digital twins. Annual Reviews in Control, 2021, *51*, 357–373.

Wang, C. Ghenniwa, H., and Shen, W. Real time distributed shop floor scheduling using an agent-based service-oriented architecture. Int. J. Prod. Res., 2008, *46*, 2433–2452.

Wang, L., Li, Q., Ding, R., and Sun, M., Wang, G. Integrated scheduling of energy supply and demand in microgrids under uncertainty: A robust multi-objective optimization approach. Energy, 2017, *130*, 1–14.

Wenzelburger, P., and Allgöwer, F. Model Predictive Control for flexible job shop scheduling in Industry 4.0. Appl. Sci., 2021, *11*, 8145.

Yang, W., and Takakuwa, S. Simulation-based dynamic shop floor scheduling for a flexible manufacturing system in industry 4.0. In: Proceedings of the Winter Simulation Conference (WSC), Las Vegas, NV, USA, 3–6 December 2017; IEEE: Piscataway, NJ, USA, 2017; pp. 3908–3916.

Yang, W., and Takakuwa, S. Simulation-based dynamic shop floor scheduling for a flexible manufacturing system in industry 40. In: Winter Simulation Conference (WSC); 2017, pp. 3908–3916. IEEE.

Zhang, X., Liu, X., Tang, S., Królczyk, G., Li, Z. Solving scheduling problems in a distributed manufacturing system using a discrete fruit fly optimization algorithm. Energies, 2019, 12, 3260.

Zhou, R., Chen, G., Yang, Z. H., Zhang, J. B. Distributed manufacturing scheduling using a novel cooperative system. In: Proceedings of the 2008 IEEE International Conference on Service Operations and Logistics, and Informatics, Beijing, China, 12–15 October 2008; IEEE: Piscataway, NJ, USA, 2008; Volume 1, pp. 256–260.

Part II

Case Studies on Collaborative Management Paradigms Applications

5

Group Decision-making Approach for Ranking and Selecting Maintenance Tasks for Joint Scheduling with Production Orders

Group decision-making has captured the attention of researchers for decades but due to its importance and complexity further explorations and studies, namely, for its application in industrial engineering continue to be needed in the current digital age. In this paper, a group decision-making approach is put forward for evaluating and selecting maintenance tasks to enable joint maintenance and production management, by using a collaborative management system. The proposed approach includes a two-stage assessment method, which enables a set of decision- makers to collaborate in ranking and selecting a set of maintenance tasks for being jointly scheduled with production orders. The group decision-making approach uses a dynamic multi-criteria decision model that aggregates information about historical, current, and provisional data about maintenance tasks. The proposed collaborative approach is illustrated through an application example and further contextualised within the state of the art. This study permits to us to realise that collaborative management approaches, namely, based on the group decision-making approach, enable conducting a dynamic, integrated, distributed, intelligent, predictive, time and condition based maintenance task management in real time, postulated on the fusion of past, present, and future data, and that there is still a lack of contributions regarding the use of collaborative approaches in industrial management.

1. Introduction

Group decision-making (GDM) is a research topic that falls within collaboration, and collaborative management domain (Varela et al., 2022; Varela et al., 2022a), and is of primer relevance in the digitalization era, by promoting and enabling a sustainable development of companies (Varela et al., 2022, 2023; Varela et al., 2022a,b, 2023).

The development of GDM approaches require the acquisition, processing, and analysis of varying kinds of data, which typically is expressed through Key Performance Indicators (KPI), for being monitored and controlled by using appropriate dashboards and systems (Simonov et al., 2018).

Over the last 20 years, the processing industries have invested heavily in automation and plant information systems such that the data is now accessible. As a result, this data should now be possible to put into productive usage. The challenge with raw data, no matter how accessible, is that it is just data, and data still requires a lot of work before it can be turned into knowledge. In most cases, the data needs to be validated, analysed, and converted into a level of knowledge that is actionable, and this can require a significant investment of time and resources.

Several kinds of KPI have been frequently used to analyse companies' performance in a given context intending to reach certain organis ational goals. Every companies' functional group defines its objectives and targets, and if the raw operational data can be converted in KPI for being processed and analysed, preferably in real-time, a better monitoring and control on the processed data can be reached and thus better decision-making processes can occur.

Information monitoring, based on proper decision support systems (DSS) is fundamental for obtaining maximum profit out of KPI through the use of suitable systems' data visualisation interfaces, and which are currently being improved by using advanced and dynamic digital dashboards, namely, through the use of power BI (business intelligence) graphics that enable real-time generated data to be analysed. Although, the real potential of a system data visualisation interface relies on its interactive ability to quickly sort and display the consolidated performance metrics in order to highlight the top priority requirements and provide guidance on further actions required. This is performed through a combination of filtering, uncertainty filtering, normalisation, weighting, aggregation, ranking, and selection techniques, and put available through appropriate collaborative systems and platforms (Campanella et al., 2012; Arrais-Castro, 2015a,b, 2018; Jassbi et al., 2016; Varela et al., 2018; Simonov et al., 2018).

As mentioned by Knoben and Oerlemans (2006), inter-organizational collaboration enables unification of disparate systems and solutions in order to achieve overall strategic and operational excellence. Therefore, intra and intercompany and manufacturing environments collaboration should be intensified, and this can be accomplished by putting into use appropriate group decision support approaches. Such approaches will permit fully integrated decision-making processes among diverse manufacturing plants and resource interactions, by using suitable platforms and systems offering effective support to carry out distributed and integrated management. Such unified workflow environments will thus promote and enable collaboration and support different decision-making teams to work together with an understanding of their specific requirements in the context of a general view over an extended and/or virtual enterprise, which is of utmost importance in manufacturing and management, for instance, in collective maintenance and production management.

Maintenance planning plays an important role in every service and manufacturing system, as it makes them more reliable and keeps them at an optimal operational level to provide high quality services and products. Additionally, the proportion of maintenance costs to the total production costs, which ranges from 15% to 70% according to the type of the manufacturing firm (McCall, 1956), makes maintenance planning a critical issue. Maintenance models can be broadly classified into two types: time-based and condition-based models (Rahmati et al., 2018). Recently, the joint optimization of production and maintenance plans has gained more attention. However, it has not been well studied compared to research on optimising maintenance planning and production schedules independently (Pan, Liao, and Xi, 2012; Bajestani et al., 2014; Fitouhi et al., 2017). In addition to the above-mentioned classification of maintenance models, integrated maintenance and production scheduling models can also be classified into two types: integrated maintenance and production scheduling models with time-based maintenance activities; and integrated maintenance and production scheduling models with condition-based maintenance activities.

Maintenance operations can be classified into two main large groups: corrective maintenance (CM) and preventive maintenance (PM). CM corresponds to the actions carried out when the failure has already taken place, and PM is the action taken on a system while it is still operating. PM is carried out in order to keep the system at the desired level of operation, and several PM policies can be defined (Rahmati et al., 2018; Sloan and Shanthikumar, 2000; Taghipour and Azimpoor, 2018), with the aim of determining when it is necessary to carry out PM operations on the machines according to different criteria used.

Besides maintenance planning, the maintenance and production scheduling is a critical decision process for the gainful management of any manufacturing system. While the first ensures reaching the production goals, besides the satisfaction of customer demands, the second ensures that manufacturing assets are available and in the proper condition to perform their required production tasks when needed. The two decision processes are interdependent since they share a clear common issue, the manufacturing assets that are used through production and restored by maintenance actions.

Integrating production and maintenance scheduling will enable optimising the joint production orders and maintenance task programming, while avoiding penalising drawbacks in companies (Ladj, A., Varnier, C., Tayeb FB-S. IPro-GA, 2016).

Although, according to the study conducted, it is possible to realis e that there is still a gap in this research domain, as insufficient work has been put forward regarding joint maintenance and production management strategies and tools.

In order to provide a contribution in this focused domain, in this paper, a GDM approach for supporting maintenance tasks assessment and selection is presented, for enabling further joint maintenance tasks and production orders scheduling, to reduce the lack of research that still prevails in this scientific domain. The proposed GDM approach is based on a Dynamic Multi-criteria Decision Model (DMCDM) (Varela et al., 2018), implemented through a two-stage maintenance tasks processing (2SMTP) methodology, which is available through a Collaborative Management System (CMS) that further permits the integrated maintenance tasks and production orders scheduling.

To properly expose the developed work, this paper follows with a resumed literature review about DSS, MCDM, and GDM, along with a general overview about approaches and systems for supporting maintenance and industrial operations management in Section 2. Next, the developed collaborative management system for joint maintenance tasks and production orders processing, along with the underlying GDM approach, and the proposed two-stage maintenance tasks assessment and selection method is briefly described in Section 3, and further illustrated through an industrial example of application in Section 4. Follows a final discussion and contextualization of this work within the state of the art in Section 5, and the main conclusion and proposed future work in Section 6.

2. Literature Review

In this section, a general overview about decision support methods and systems, along with GDM approaches is briefly presented next, in subsection 2.1, followed by a summaris ed description of maintenance and industrial operations management approaches and DSS in subsection 2.2.

2.1 Decision-making Methods and Systems for Group Decision-making Support

A DSS can be explained as an interactive computer-based system, which can be helpful for decision- makers to use quantitative models and data for solving complex problems (Bhatt and Zaveri, 2002; Lee and Huh, 2006). A DSS enables supporting more or less complex decision processes by using different kinds of middleware and technology, and tools (Sprague and Carlson, 1982; Zarate, 1991; Vieira, et al., 2018; Vafaei, et al. 2019). Keenan (2016) mentioned that DSS have been developed since the 1970s, and since then continued growing and improving, based on new technologies, namely, about databases and visual interfaces applied for properly supporting decision-making processes. DSS mostly involve Management Science and Operations Research fields. DSS and management strategies have thus a meaningful relationship in manufacturing environments for reaching well-suited decisions (Brannback, 1994).

During recent decades, DSS have been developed in different contexts, and some contributions are summaris ed next.

Group Decision Support Systems (GDSS) and Executive Information Systems, which was changed to the Enterprise Information Systems (EIS), introduced to support DSS tools are becoming much improved and more effective. GDSS currently provide many useful options, including brainstorming, idea assessment, and some other facilities for enabling communication in more or less complex problem solving scenarios (Costa et al., 2003; Limayem and Banerjee, 2006; Varela, et al., 2021), along with other kinds of the so-called Integrated Decision Support System (IDSS) that enable improving the effectiveness of classical DSS by combining them (Liu et al., 2010).

More recently, DSS has been applied in integrated models with Multiple Attribute Decision Making (MADM) and Multiple Objective Decision Making (MODM) in a general framework of Multiple Criteria Decision Making (MCDM) for endowing a better process and environment in decision support (Jaramillo et al., 2005; Qureshi et al., 2017). Bakshi et al. (2015) mentioned that when there is uncertainty in decision-making processes the MCDM models will become more complicated thus requiring appropriate Multi-Criteria Decision Support Systems to present appropriate solutions in practice. The authors mention a new DSS established based on models, survey (literature review), and human experts interacting through a proposed framework. The main issue of their research was selecting the main criteria in MCDM models. Some other studies applied this kind of approach in practice, and some are resumed next, to mention a few.

Taha and Rostam (2012) applied a hybrid fuzzy AHP-PROMETHEE as the main part of a DSS for machine tool selection in a flexible manufacturing cell. They mentioned that their research shows that MCDM methods can be a useful part of a DSS and that their vision would be helpful in decision-making in solving complex cases.

Razmak and Aouni (2015) reviewed research related to MCDA and DSS and found more than 100 research articles for analysis. They categoris ed the articles into nine different sections, regarding their application fields which. were: Production and Supply Chain Management; Education; Human Resource Management; Finance and Investments; Real Estate and Constructions; Environmental aspects; Medical aspects; Electronic business and electronic commerce, and Multimedia.

Leyva Lopez et al. (2016) proposed a model and system for supporting group decision-making based on a MCDM approach. The authors state that their approach was structured and based on the ELECTRE method and designed completely based on the web to make the underlying process more reachable and easier applicable in practice. Their proposed GDSS enables them to put forward some advice for decision makers in order to help them manage their priorities and preferences to allow proper decision rules with some degree of consistency and consensus.

In other works (Arrais-Castro et al., 2018; Simonov et al., 2018; Varela et al., 2018), DSS models were proposed by using different kinds of approaches, in various application contexts. According to the examples provided, it is possible to realise that DSS and approaches are applied in various contexts and manufacturing and management environments, thus there is still a need for new contributions to increase its full practical capability and usability, for instance, in the industrial context. Furthermore, decision-making, with uncertainty treatment and future or prospected data processing, needs integrated and advanced DSS models and systems to continue being developed to decrease ambiguity and vagueness of knowledge about forecasted data, which has become, especially currently, in the digital age, more urgent and necessary, for putting into practical use in manufacturing management (Putnik et al., 2021).

2.2 *Approaches and Systems for Supporting Maintenance and Industrial operations Management*

Maintenance is a crucial activity in industry, with a significant impact on costs and reliability, being immensely influential on a company's ability to be innovative, while permitting costs reduction and global benefits, namely, increased quality and general performance.

In the scope of maintenance management, any unplanned downtime of machinery equipment or devices usually degrades or harms a company's core business, potentially resulting in significant penalties and unmeasurable reputation loss. According to some studies, operation and maintenance costs can range from 15% to 70% of total production cost in some companies (Bevilacqua and Braglia, 2000; Gong and Qiao, 2014). Therefore, it is critical for companies to develop a well-implemented and efficient maintenance strategy to prevent unexpected drawbacks, and improve overall reliability, while reducing manufacturing systems' operating and maintenance costs.

The evolution of modern techniques, namely with the emergence of the Internet of things (IoT), along with varying kind of sensing technology, and new or improved artificial intelligence approaches and tools, among others, stimulates a transition of maintenance strategies from Reactive Maintenance (RM) to Preventive Maintenance (PM), and to Predictive Maintenance (PdM) (Jimenez et al., 2020). RM is only executed to restore the operating state of the equipment after failure occurs, and thus tends to cause serious unproductive times, while frequently resulting in high response and reparation costs. PM is carried out according to a planned schedule based on time or process iterations to prevent breakdown, and thus may perform unnecessary maintenance, typically resulting in high prevention costs. To achieve the best trade-off between the RM and PM, the PdM can be performed, based on some online assessment of the condition of manufacturing assets, and thus reach timely interventions before failure occurs, while preventing high maintenance frequency, unplanned RM, and the incurrence in increased costs associated to frequent PM.

Asset management deals with the optimization of manufacturing assets used for reducing costs. An asset management system manages the assets over the whole life cycle, especially their reliability and efficiency. It is also responsible for optimising utilisation and cost-effective maintenance of the assets. Moreover, it generates and provides information regarding the so-called "asset health" development and prognosis to support decision-making of the enterprises' production management (Namur 2009). Using the "asset health" information to generate an optimal production plan is a viable solution to better integrate a maintenance and a production planning system to increase the overall performance (e.g., in terms of costs) of manufacturing operations. Although some work was already carried out in this sense, industry is still lacking appropriate and effective systems for supporting advanced maintenance and production management (Zhai et al., 2021).

Biondi and Harjunkoski (2017) proposed a joint scheduling approach for the production and maintenance of process plants that explicitly keeps track of the assets' life cycle. The scheduling system includes a simple model of the asset wear that can be based on the concept of residual useful life (RUL) or of probability of failure. The authors state that the asset monitoring system is responsible for providing two types of information to the scheduling system: on the one hand, an estimation of the parameters describing the wear caused by the production on the asset. On the other hand, if an extraordinary condition of the asset is detected, it is responsible for updating a current RUL in the asset wear model of the scheduling system. Assets' health information, along with the production orders, is managed by the scheduling system that takes care of the sequencing and timing of production tasks on the plant and triggers a maintenance action on the assets whenever this is required. According to the authors, their proposed method makes an effective use of factory units' health information to generate a feasible plan for joint production and maintenance planning (Biondi and Harjunkoski, 2017).

Based on Staufen (2018), PM has not been properly explored in the industry. A survey in 2020 shows that PM continues to be a hot topic, for example, to determine the best point in time to do maintenance tasks (Zhai et al., 2020).

Two types of flexible PM strategies, i.e., time-based PM (TBPM) and condition-based PM (CBPM), are commonly analysed and applied (Wang, Yan, and Zhang, 2021). According to these authors, the application of TBPM is straightforward and relative ease of implementation, however, TBPM may lead to under-or over-maintenance due to inaccurate estimates of the stability of production systems. In contrast, CBPM is of more complexity, which continuously monitors and analyses the machine status to determine the implementation of the maintenance activity. The authors state that despite the complexity of computational requirements and uneven maintenance cycles, CBPM strategy can reduce the maintenance frequency to a minimum necessary level, thus improving a global production system's productivity level.

Some examples of application of TBPM in diverse kinds of production scenarios, integrating different production scheduling strategies, are presented by several researchers (Chen, 2000; Chen et al., 2006; Mosheiov and Sarig, 2009; Yang et al., 2011), while CBPM has also been focused by several other researchers, for instance (Zandieh et al., 2017; Rahmati et al., 2018; Sloan and Shanthikumar, 2000; Ghaleb et al., 2020), just to mention a few.

Prognostics and health management (PHM) is a relatively young engineering discipline that aims to enable "real-time health assessment of a system under its actual operating conditions as well as the prediction of its future state based on up-to-date information" (Kim N-H, An D, Choi J-H, 2017), with PdM being the underlying maintenance strategy that uses prognostics results of PHM.

(Li et al., 2019) state that varying operational conditions have two major effects on system degradation: Firstly, varying operational conditions influence the speed of degradation. Secondly, they lead to sudden signal changes and change points, which result in high variance of raw sensor readings. Thus, varying operational conditions pose an obstacle to prognostics (Zhang et al., 2020) and are considered to be a focal point for modern PdM modelling (Aydemir and Acar, 2018).

According to Assaf, Scarf, and Iung (2019), prognostics incorporates three tasks: ''State estimation'' (estimate the current health or degradation state of the system based on historical data), ''State prediction'' (predict the health or degradation state for future periods based on historical data), 'EoL' (''End of Life'') or ''RUL prediction'': Determine the RUL before failure or before exceeding the failure threshold for some identified degradation behaviour. The author highlights that RUL can refer to actual failure or remaining time until certain quality requirements of a product cannot be met.

Databased RUL prediction can be formulated as a supervised (Aggarwal et al., 2018) or a semi-supervised machine learning (ML) problem (Yoon et al., 2017). According to these authors, the high amount of required failure data to derive RUL labels for supervised prediction models is often not available in industrial practice.

Health prognostic approaches in PHM are commonly classified into physics-based, knowledge-based, and data-driven approaches (Bektas et al., 2019). Physics-based models describe the phenomena of failure and degradation as physical or mathematical "white box" models. Although physics-based models can achieve high accuracy, their development is usually costly (Bektas et al., 2019). Knowledge-based models collect identified degradation behaviours and failure events in a historic database and assess the similarity of a currently observed system state with the entries of a knowledge base (Sikorska et al., 2011). Data-based approaches make use of the system condition monitoring (CM) data to derive transparency of the system health state and predict the RUL (Song et al., 2018; Jia et al., 2018; Wang et al., 2017), further enabling to assess the uncertainty of the prediction (Benker et al., 2020). Data-based methods encourage the use of highly adaptable ML, including deep learning (DL) algorithms (Zhang et al., 2018), in scenarios where large amounts of condition monitoring data are available and the system operation is subject to variations, partially unknown conditions or a variety of failure modes.

For an overview of knowledge-based approaches, as well as advantages and limitations of data- and knowledge-based approaches, the reader is referred to Ran et al., (2019), where a survey of predictive maintenance systems purposes and approaches is presented. Next, some additional work is briefly referred to.

The authors in Malhotra et al. (2016) propose an approach for combined health indicator (HI) estimation and RUL prediction. The publications by Wang (2010) and Wang et al. (2008) are among the first research works to explicitly consider the effects of time-varying operating conditions on system degradation analysis.

Li et al. (2019) model a dynamic, operation-specific degradation rate as a state transition function based on Wiener process and time-scale transformations, which capture the effect of operating conditions on the degradation curve. A measurement function smoothens the jumps in the degradation signal at operation condition change points by mapping each condition to a condition-specific baseline. The approach proposed by the authors is evaluated on a simulated data set of bearings, which are subject to varying rotational speeds, as well as on a data set from an accelerated degradation experimental study of rolling element bearings.

Luo et al. (2019) propose a DL approach for health estimation and fault detection of CNC machine tools operating under time-varying conditions. In the first step, the authors use a DL model composed of stacked auto encoders (AE) and a feed forward neural network to extract impulse responses from vibrational CM data. The training and test data sets for the DL model are prepared manually by labelling whether randomly selected time windows contain an impulse response or not. In case of an impulse response, the vibration signal represents the reaction of the system to sudden forces and impacts during time-varying machining processes. After training, the DL model is used to automatically identify impulse responses in the CM data. Subsequently, the first four natural frequencies and the damping reactions of the machine tools are extracted from two different impulse responses representing two different working conditions. The authors find that the natural frequencies barely change with varying operational conditions and thus are a robust feature for HI construction. The HI is computed as the cosine similarity in the space of extracted dynamic features comparing current observations with an initial vector representing the normal state. According to the authors, since the HI is based on operation-condition invariant features, the HI is robust to different working conditions. However, the approach is not capable of performing an operation-specific prediction of system health for future loads. In contrast to most other research, the approach was evaluated on a real industrial data set, composed of vibration signals from 288 days of industrial operation.

Michau and Fink (2019) propose an unsupervised approach for system monitoring in a setting where a fleet of similar safety-critical systems is to be monitored over time. The training data for a specific system instance is enhanced by CM data from other instances of the fleet to enable CM early in a system's operational life. The authors use a variational auto encoder (VAE) architecture to model a shared latent space for the fleet, which is trained in an adversarial manner. A new loss function is designed to preserve instance-specific behaviours in the shared latent space. The health prediction is framed as a one-class classification, which aims at predicting whether the CM data is faulty or healthy. The method is evaluated using a real data set from a fleet of 112 power plants operated in different geographical locations and under different operational conditions. The authors say that their results show that the shared latent representation and feature alignment yield an efficient and unsupervised feature representation in a setting of complex systems subject to varying conditions, which is useful for downstream PHM modelling.

The integrated optimization of production scheduling and machine maintenance has been known as a complex combinatorial optimization problem, in which heuristic or meta-heuristic approaches are commonly employed aiming to find some satisfied solutions in a short time. With the advent of artificial intelligence and ML and DL, the application of scheduling rules-based reinforcement learning (RL) to the field of scheduling has become possible (Wang and Usher, 2005). However, little empirical research concerning the application of RL to integrated decision making of production scheduling and machine maintenance has been conducted (S. Zhai, B. Gehring, G. Reinhart, 2021).

In S. Zhai, B. Gehring, G. Reinhart (2021), a machine degradation modelling under varying operational conditions, enabling subsequent integrated scheduling of maintenance and production ("PdM-integrated production scheduling": PdM-IPS) is introduced. The underlying model is a conditional variational auto encoder (CVAE) that is used for calculating and quantifying the change of the machine health condition after producing specific product sequences.

The gap that continues existing regarding contributions of integrated maintenance and production management approaches and systems motivated this work, in order to contribute to this scientific domain, and the proposed collaborative management system, based on a group decision-making approach is briefly described and illustrated in the next sections.

3. Collaborative Management System Based on a Group Decision-making Approach

Collaborative management is of utmost importance in the current digital age, enabling and promoting a sustainable development of companies (Varela, Putnik, and Romero, 2022, 2023; Varela et al., 2022a,b, 2023). In this paper, a group decision-making architecture is proposed to enable collaborative management, and Figure 1 shows an example of its application in an industrial company that includes three work centres (WC1, WC2, and WC3), which interact with each other and with the main company's factory, through its underlying brokering service, besides communicating with clients, and maintenance technicians.

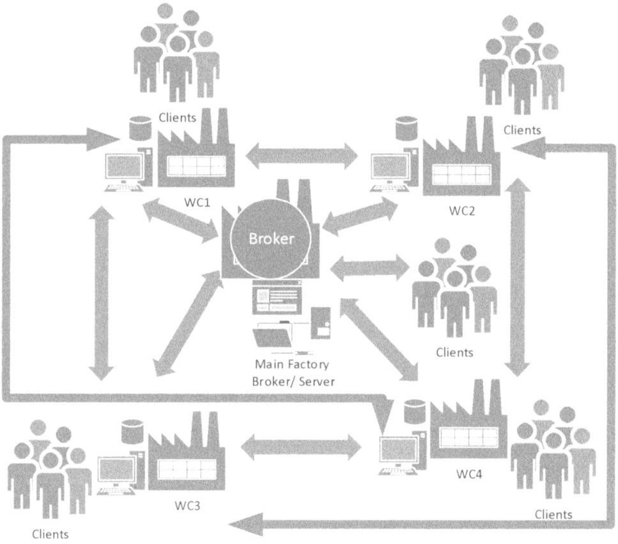

Figure 1. Group decision-making architecture.

A CMS underlying the proposed GDM architecture was developed to enable intra and inter factories and/or work centres collaboration for jointly reaching integrated maintenance tasks and production operations scheduling, and an interface of the CMS is shown in Figure 2, an interface for processing a data fusion function of the DMCDM used in this work that will be further explained through an application example.

This CMS enables a wide range of diverse management functions in industrial management, namely, underlying the proposed GDM approach, which is carried out by using a maintenance tasks processing methodology with three phases, based on a two-stage assessment method, which makes use of a DMCDM, as expressed in the Figure 3.

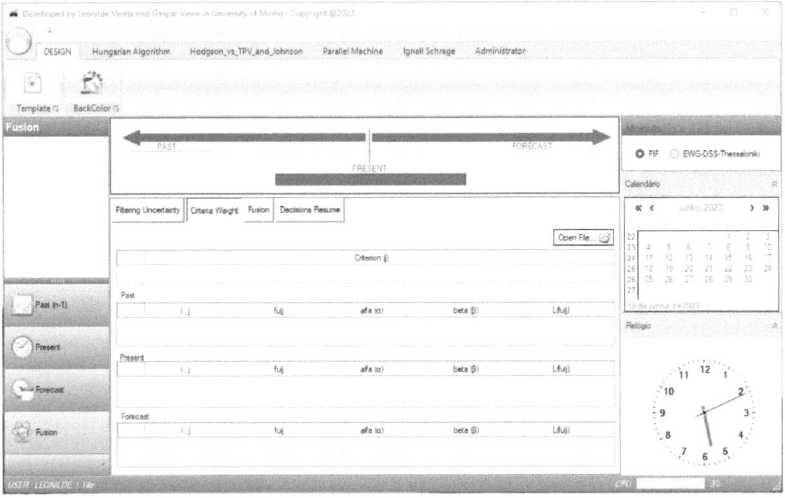

Figure 2. Collaborative management system's interface illustration: data fusion function.

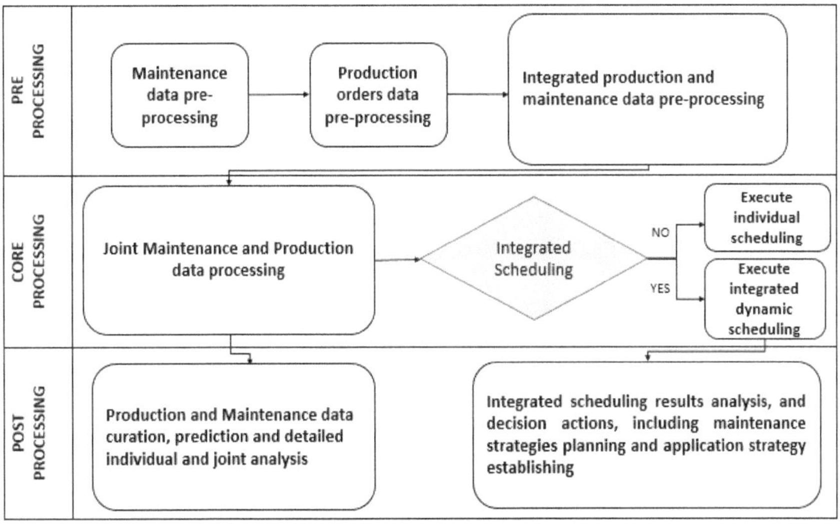

Figure 3. Proposed maintenance tasks processing methodology based on a two-stage assessment method.

The two-stage assessment method based on DMCDM was used in an industrial company, in the scope of a research project to enable the joint processing of maintenance and production orders information, and a case study is briefly described next.

4. Application Example

4.1 Maintenance Tasks Assessment Methodology Based on a DMDCM

The maintenance tasks assessment methodology used in this work uses a DMCDM (Jassbi et al., 2013; Varela et al., 2018), by including two stages, for intra- and inter-work centres tasks' evaluation and selection.

1st Stage) Intra-work centres evaluation: Includes 6 steps: normalis ing/ fuzzifying, weighting, uncertainty filtering, and data aggregation or fusion, for ranking and selecting the maintenance tasks (Varela, Arrais-Castro, and Ribeiro, 2018).

Step 1) Data acquisition and matrices construction: First, the definition of the evaluation criteria for processing the data about 3 moments: past, present, and future, have to be defined, and Figures 4 to 6 show an example for the company's WC1, by using different kinds of criteria, about: maintenance cost (MC), Lack of Production Quality (LPQ), Overall Equipment Effectiveness (OEE), Lack of Safety Indicator (LSI), Mean Time Between Failures (MTBF), Mean Time To Repair (MTTR), and Downtime (DT), for processing past and future maintenance tasks' data; and MC, along with Service Time (ST), and Lead Time (LT), for processing current maintenance tasks' data, which were applied for ranking a set of six maintenance tasks (M1_1 to M1_6) of WC1. The current or present data is acquired, in real-time from the shop floor, by using appropriate communication means and devices (Vieira et al., 2018; Varela et al., 2021).

Maint. Task	Maint. Cost MC	Lack of Production Quality LPQ	Overall Equipment Effectiveness (100%) OEE	Lack of Safety Indioator LSI	Mean Time Between Failures MTBF	Mean Time To Repair MTTR	Downtime DT
MT1_1	630,00	28%	88	18%	28000	2	5
MT1_2	398,50	33%	79	14%	17500	4	4
MT1_3	400,00	17%	90	11%	18900	3	3
MT1_4	730,00	20%	85	13%	22400	2	3
MT1_5	490,00	15%	83	15%	15600	5	5
MT1_6	330,00	12%	92	12%	20580	1	4

Figure 4. Past data matrix (Varela et al. 2024).

Maint. Task	Maintenance Cost MC	Service Time ST	Lead Time LT	Extended ST (EST) ST + (max(LT-d),0)
MT1_1	635,00	100	5	105
MT1_2	405,00	120	2	122
MT1_3	410,00	85	4	89
MT1_4	650,00	90	3	93
MT1_5	450,00	100	4	104
MT1_6	335,00	115	3	118

Figure 5. Present data matrix (Varela et al. 2024).

Maint. Task	Maint. Cost MC	Overal Equipment Effectivness OEE	Downtime DT	Production Quality PQ	Mean Time To Repair MTTR	Mean Time Between Failures MTBF	Safety Indicator SI
MT1_1	640,00	85%	20	86%	5	23500	4
MT1_2	498,50	90%	35	92%	10	13000	4
MT1_3	410,00	70%	17	85%	6	25500	3
MT1_4	700,00	75%	10	73%	7	15000	3
MT1_5	395,00	73%	20	80%	9	14000	5
MT1_6	340,00	84%	18	90%	4	11000	4

Figure 6. Future data matrix (Varela et al. 2024).

The future data can be obtained by applying some forecasting method, namely, by using some ML approach (Putnik et al. 2021) or be based on known real or estimated data. Prediction may also be performed using expert judgement or quantitative methods (forecasting), such as moving linear averages, quadratic averages, and other techniques.

Step 2) Normali sation/fuzzification

In the second step, a normalisation/ fuzzification process underlying the DMCDM (Varela, Arrais-Castro, and Ribeiro, 2018) was performed (Figure 7), to process imprecision by using fuzzy logic for criterion evaluation. Normalisation guarantees that values are numerical and comparable, simple triangular membership functions were used to represent the acceptable criterion values, as all expected criteria fit in the "lower is better" and "higher is better" categories (Varela and Ribeiro, 2003). This process is essential to enable values aggregation, and the simplest method consists on dividing a value by the maximum existing one in the set (when high values are favourable to the decision) or by the minimum (when low values are favourable, such as a cost) (Jassbi et al., 2014).

Step 3) Uncertainty filtering

In order to filter uncertainty, a method underlying the DMCDM referred to in Varela, Arrais-Castro, and Ribeiro (2018) is used, which considers two

MC

	x	u(x)
1,1	630,00	0,250
2,1	398,50	0,829
3,1	400,00	0,825
4,1	730,00	0,000
5,1	490,00	0,600
6,1	330,00	1,000

bi	330	
pi	400	(730-330)

Figure 7. Normalisation and fuzzification example for the MC criterion (Varela et al. 2024).

parameters, accuracy and confidence to ' filter' the membership function values. The accuracy parameter expresses deviations from nominal values and the confidence expresses the degree of trust on the data gathered.

The logic of this filtering process is that if we do not trust an input source (e.g., confidence on data is only 80%) then the initial value must decrease proportionally (e.g., a value 10 would be reduced to 8), Thus accommodating deviations in the value, for example, +3 or –3 from a value of 10.

Let a_{ij} be the accuracy associated with criterion *j* for MT*i*, representing a left or right deviation from the original value; when a_{ij} is zero it means we accept the gathered value without deviation errors.

The confidence, wc_j, is a percentage, as for example, we trust with 90% the values for "Maintenance Cost, MC".

Additionally, $\lambda \in [0,1]$, is a parameter that reflects the decision- maker's attitude. Values close to zero indicate an optimistic attitude; higher values indicate a pessimist attitude.

The accuracy rate, expressing the allowed deviation from the base values, is defined for each criterion, based on the associated data quality. The value also reflects the imprecision associated with the data gathering process. Based on the criteria and its associated confidence rates, the filtered imprecision values, fu_{ij} (e.g., acij), were calculated, as illustrated next for the MC criterion.

Hence, the adjusted membership value is calculated using the following formula (Varela et al., 2018):

$$fu_{ij} = wc_j * (1 - \lambda * \max_{x \in [a,b]} \{|\mu(x) - \mu(x_{ij})|\} * \mu(x_{ij})) \quad (1)$$

Where [a,b] is the inaccuracy interval:

$$a = \begin{cases} min\,(D), & if\ x_{ij} - a_{ij} \leq min\,(D) \\ x_{ij} - a_{ij}, & if\ x_{ij} - a_{ij} > min\,(D) \end{cases} \quad (2)$$

$$b = \begin{cases} x_{ij} + a_{ij}, & if\ x_{ij} + a_{ij} \leq max\,(D) \\ max\,(D), & if\ x_{ij} + a_{ij} > max\,(D) \end{cases} \quad (3)$$

Using the function (1), along with (2) and (3), we are able to penalis e input values, which display any of the two types of uncertainty, i.e., inaccuracies or lack of confidence on data, within an optimist or pessimist view from the decision maker.

MC

wc_j	100%	λj	1	(in)acuracy int.	0%	pi	400	
i,j	u(xij)	xij	aij	a	u(a)	b	u(b)	acij
1,1	0,250	630	0	630	0,250	630	0,250	0,250
2,1	0,829	399	0	399	0,829	399	0,829	0,829
3,1	0,825	400	0	400	0,825	400	0,825	0,825
4,1	0,000	730	0	730	0,000	730	0,000	0,000
5,1	0,600	490	0	490	0,600	490	0,600	0,600
6,1	1,000	330	0	330	1,000	330	1,000	1,000

Figure 8. Uncertainty filtering example for the MC criterion (Varela et al. 2024).

Step 4) Weighting

Step 4 enables us to allow different weights for different temporal stages or criterion. Here we will use linear weighting functions to express the relative importance of criteria. These functions allow penalis ing or rewarding bad or good levels of criteria satisfaction, i.e., instead of assigning single weights, we represent them using a function that depends on criteria satisfaction (equation 4):

$$L\,(fu_{ij}) = \alpha \,*\, \frac{1+\,\beta\, fu_{ij}}{1+\,\beta}\;,\; 0 \le \alpha,\, \beta \le 1 \qquad (4)$$

where α defines the semantic importance of criteria ('1' –very important, … '0' -ignored), and the β parameter defines the slope for the weighting function (a higher value or slope means a steeper function, thus a higher penalty, e.g., '1', and '0' –null penalization) to penalise, more or less, badly satisfied criteria. For example, if we assign to criterion Maintenance Cost, MC the values α=1 and β=0.67, we are defining this cost as a "very important" evaluation parameter with an average slope decrease. In this case, we want to reward the best quotes and penalise the bad ones (i.e., we want to reward lower costs).

j=1 (MC)

i,j	fuij	alfa	beta	L(fuij)
1,1	0,250	1	0,670	0,699
2,1	0,829	1	0,670	0,931
3,1	0,825	1	0,670	0,930
4,1	0,000	1	0,670	0,599
5,1	0,600	1	0,670	0,840
6,1	1,000	1	0,670	1,000

Figure 9. Weighting example for the MC criterion (Varela et al. 2024).

Step 5) Aggregation

After the four previous steps, we have a weighted vector for each criterion. Step 5 is to determine the score (rating) for each time period, i.e., past, current, and future, by using an approach that is illustrated for the past values about the MC criterion. The following results were obtained for historic information, using the data fusion (equation 5):

i	MC	OEE	DT	PQI	MTTR	MTBF	SI
1	0,037	0,083	0,026	0,142	0,000	0,020	0,177
2	0,145	0,111	0,166	0,041	0,074	0,057	0,062
3	0,170	0,022	0,013	0,000	0,072	0,043	0,000
4	0,000	0,040	0,056	0,032	0,041	0,022	0,000
5	0,100	0,011	0,067	0,061	0,107	0,088	0,167
6	0,220	0,000	0,000	0,014	0,055	0,000	0,073

i	r_i	Maint. Task
1	0,485	MT1_1
2	0,656	MT1_2
3	0,320	MT1_3
4	0,191	MT1_4
5	0,601	MT1_5
6	0,362	MT1_6

Figure 10. Aggregation example for the MC criterion in the past data matrix (Varela et al. 2024).

$$r_i = sum\left(\frac{L\left(fu_{ij}\right)}{\sum_{k=1}^{n} L\left(fu_{ij}\right)} * fu_{ij}\right) \quad (5)$$

Step 6) Decision

Once applying the steps underlying the DMCDM: normalis ation/ fuzzification, weighing, uncertainty filtering, and aggregation or data

Maint. Task	Score	Position
MT1_2	0,65644141	1
MT1_5	0,6009151	2
MT1_1	0,48491111	3
MT1_6	0,36222612	4
MT1_3	0,31985451	5
MT1_4	0,19085262	6

Figure 11. Decision matrices example for the MC criterion (Varela et al. 2024).

fusion to the past information of the WC1, it is possible to obtain the following rankings of the corresponding 6 maintenance tasks considered in this example:

Next, we repeat the process underlying the DMCDM for future information, and in this case study the same criteria that have been used for past information evaluation that were used for future data processing. Once having calculated the historical and prediction (future) scores for each alternative, we also need to evaluate the present status (present data).

Evaluating the present or current data means to evaluate the proposals/quotes that have been received and then fusion the respective information. For that purpose, the following criteria were used to evaluate present data: MC (Maintenance Cost), ST (Service Time), and LT (Lead Time), as previously shown.

WC1 - past

Maint. Task	Score	Position
MT1_2	0,65644141	1
MT1_5	0,6009151	2
MT1_1	0,48491111	3
MT1_6	0,36222612	4
MT1_3	0,31985451	5
MT1_4	0,19085262	6

WC1 - future

Maint. Task	Score	Position
MT1_6	0,26894646	1
MT1_3	0,20261408	2
MT1_2	0,19190658	3
MT1_5	0,18941079	4
MT1_1	0,17917192	5
MT1_4	0,08739849	6

WC1 - present

Maint. Task	Score	Position
MT1_3	0,87521921	1
MT1_6	0,70014439	2
MT1_5	0,64948589	3
MT1_4	0,59392525	4
MT1_2	0,50540424	5
MT1_1	0,30551333	6

WC1 – final scores

Maint. Task	Score	Position
MT1_3	0,5652022	1
MT1_5	0,5343635	2
MT1_6	0,4947301	3
MT1_2	0,4929897	4
MT1_4	0,3549720	5
MT1_1	0,3428058	6

Figure 12. Past, future, present, and final scores matrix examples for all the criteria in WC1 (Varela et al. 2024).

Summarising, the final ratings of the maintenance tasks regarding past, future, and present data, for the WC1, along with the final ratings, after final data weighting and fusion results for the WC1 are presented in Figure 12.

After the application of the same procedure that has been used for processing the information related to WC1 to the other two work centres (WC2 and WC3), by accomplishing the same main 5 steps of the DMCDM, the following final maintenance tasks' rankings have been obtained for these WC2 and WC3 (Figure 13).

WC2 – final scores

Maint. Task	Score	Position
MT2_2	0,5699688	1
MT2_6	0,5284007	2
MT2_5	0,4882188	3
MT2_4	0,3952934	4
MT2_3	0,3838716	5
MT2_1	0,3742694	6

WC3 – final scores

Maint. Task	Score	Position
MT3_1	0,6617229	1
MT3_2	0,5652877	2
MT3_5	0,4242146	3
MT3_3	0,4114283	4
MT3_6	0,3966577	5
MT3_4	0,3958268	6

Figure 13. Final scores matrix examples for all criteria underlying past, present, and future fused data about WC2 and WC3 (Varela et al. 2024).

Next, the two maintenance tasks, out of each WC, with the higher ratings shown next are selected for further processing in the 2nd stage of the maintenance data processing method.

It is important to note that despite M1_3 not having good rankings in terms of historical data evaluation, it benefits from the greater importance that has been given in WC1 to the present or current data.

Although, regarding the MT1_5, it reaches a higher rating than MT1_6, besides being a little worse positioned in terms of present data ratings, and with considerably worse position regarding future data, as the past data has an higher impact in the final rating than the provisional of future data, which in this case this favours MT1_5.

2nd stage) Inter-work centres evaluation

In the 2nd stage, the DMCDN is repeated for the best rankings obtained in the 1st stage. Thus, follows the application of the same approach to the six maintenance tasks from the 1st stage with a higher ranking to be further processed based on the application of the same DMCDM by the whole set of decision- makers underlying the WC1, WC2, and WC3, to obtain the final list of the three maintenance tasks with higher priority for being jointly scheduled with the production orders, by repeating the application of the same main 5 steps that were previously applied on each WC.

In this 2nd stage of the method, a higher importance has been given to the past data, followed by present and less importance to the future data, to obtain the final overall rankings.

Thus, the 3 maintenance tasks with better ratings, out of the set of the six maintenance tasks list including the two of each WC with a higher

(a) Final scores of MTi from 1st stage

i	Score	Maint. Task
1	0,5652022	MT1_3
2	0,5343635	MT1_5
3	0,5699688	MT2_2
4	0,5284007	MT2_6
5	0,6617229	MT3_1
6	0,5652877	MT3_2

(b) Final rankings of MTi from 2nd stage

Maint. Task	Score	Position
MT2_2	0,6463834	1
MT3_2	0,5581164	2
MT1_5	0,4785514	3
MT3_1	0,4774147	4
MT2_6	0,4337532	5
MT1_3	0,1708652	6

Figure 14. (a) Aggregated final scores' matrices about WC1, WC2, and WC2 from the application of the 1st stage and (b) the 2nd stage of the maintenance tasks assessment methodology (Varela et al. 2024).

priority, that were reached for being jointly scheduled with the production orders are the following: MT2_2 (being redefined as simply M2, the M3_2, redefined as M3, and M1_5, redefined as M1).

It is important to notice that eventually other criteria and importance could be defined for accomplishing this second stage of the decision method.

4.2 Collaborative Scheduling: Joint selected maintenance tasks and production orders programming

The joint collaborative scheduling is performed next, based on the model presented in Varela, et al. (2022b), to jointly program a current set of companies' production orders, along with the previously selected set of the three maintenance tasks with higher scores: MT2_2 that will now be defined simply as M2, MT3_2 as M3, and M1_5 as M1, related to the

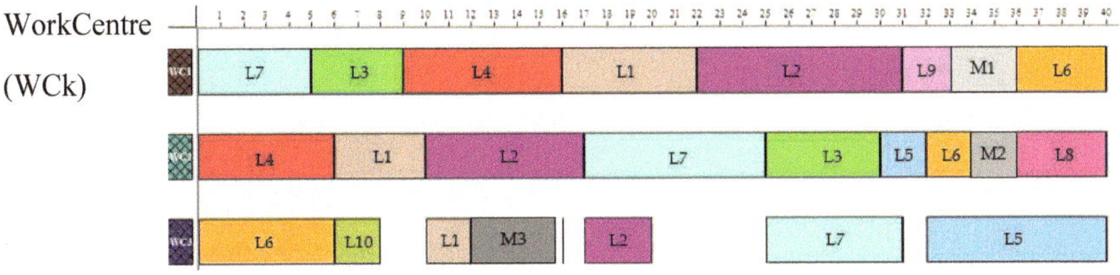

Figure 15. Gantt charts about the best solution found for scenario 1 (about the minimization of the internal performance measure, makespan, Cmax) (adapted from: Varela et al., 2022b).

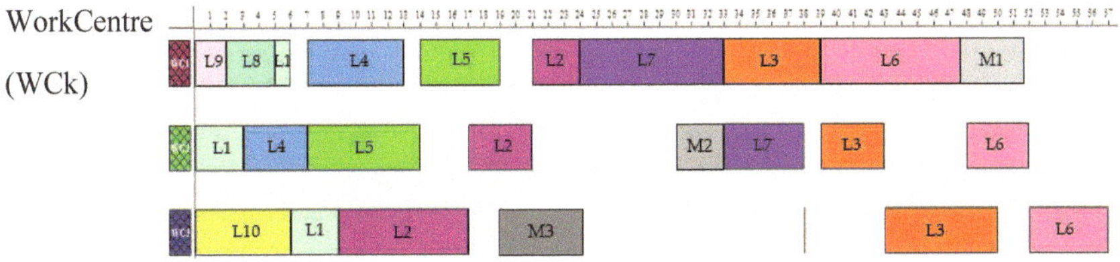

Figure 16. Gantt charts about the best solution found for scenario 2 (about the minimization of external measures, tardy jobs, Nt, and maximum tardiness, Tmax) (adapted from: Varela et al., 2022b).

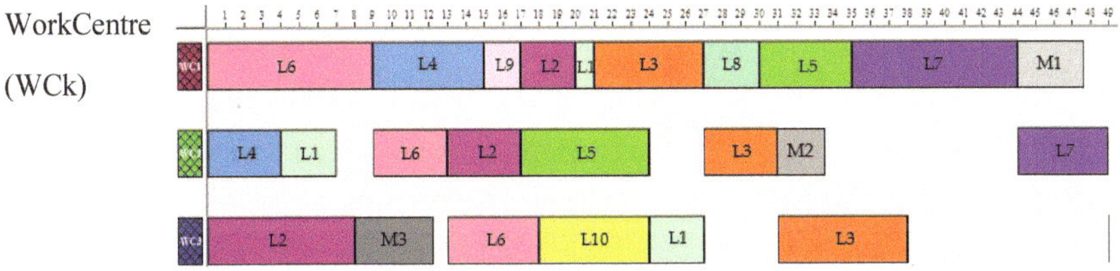

Figure 17. Gantt charts about the best solution found for scenario 3 (about the combined (50%–50%) minimization of both kinds of measures, Cmax, and Nt) (adapted from: Varela, et al., 2022b).

workcentres WC1, WC2, and WC3, correspondingly, and alternative possible solutions are shown in the Figures 15 to 17 below.

These Gantt charts express possible alternative solutions for jointly scheduling the maintenance tasks and a set of 10 lots of production orders

(L1 to L10), based on the preference that is given by the decision-making team regarding internal oriented performance measures (makespan) (Figure 15) or external oriented ones (tardiness and tardy tasks) (Figure 16) or a combination of internal and external measures (makespan and tardy tasks) (Figure 17). Thus, the developed CMS provides additional flexibility by enabling to choose the best suited application scenario, by using appropriate scheduling algorithms available for processing the joint maintenance and production tasks, according to a given industrial context and management preferences or goals of the decision making team.

5. Final Discussion

According to a study conducted, and by analysing a set of 20 publications about maintenance and production management, a resume of main contributions from the literature were analysed, considering a set of seven main dimensions underlying this study which are: dynamic, integrated, real-time, distributed, and predictive management strategies (Varela et al., 2023), along with time- and condition-based maintenance, as synthetized in the Table 1.

Table 1. Resume of main dimensions of literature contributions and proposed approach (adapted from (Varela et al. 2024)).

Dimension / Contribution	Dynamic	Integrated	Real time based	Distributed	Predictive	Time based	Condition based	Sum
(Aggarwal, et al., 2018)	X		X		X	X	X	5
(Assaf, Scarf, & Iung, 2019)	X					X	X	3
(Aydemir, Acar, 2020)	X		X		X		X	4
(Bektas, Marshall, & Jones, 2020)	X		X		X			3
(Benker, et al., 2021)			X		X			2
(Biondi, &Harjunkoski, 2017)	X	X	X		X	X	X	6
(Lee, & Chen, 2000)		X				X		2
(Ghaleb, Taghipour, Sharifi, & Zolfagharinia, 2020)		X				X	X	3
(Kim N-H, An D, & Choi J-H, 2017)			X		X	X	X	4

Table 1. contd. ...

Table 1. contd. …

Dimension / Contribution	Dynamic	Integrated	Real time based	Distributed	Predictive	Time based	Condition based	Sum
(Li, et al., 2019)	X		X		X	X	X	5
(Luo, et al., 2019)	X		X		X	X	X	5
(Malhotra, et al., 2016)			X		X	X	X	4
(Michau & Fink, 2019)	X		X	X	X	X	X	6
(Mosheiov, & Sarig, 2009)		X				X		2
(Rahmati, Ahmadi, & Govindan, 2018)	X	X	X				X	4
(Sloan, & Shanthikumar, 2000)		X				X	X	3
(Wang, & Yu, 2010)		X				X		2
(Yang, Ma, Xu, & Yang, 2011)	X	X				X		3
(Zandieh, Khatami, & Rahmati, 2017)		X					X	2
(Zhai, B. Gehring, & Reinhart, 2021)	X	X	X		X	X	X	6
This work	X	X	X	X	X	X	X	7

The analysed publications listed in the Table 1 show that, on average, three to four of the dimensions proposed for carrying out the collaborative maintenance and production management are considered. Therefore, it is noticeable that this work is novel and that there is still a gap regarding these kinds of contributions in the focused scientific and technological domain.

6. Conclusion

In this paper, a GDM approach for maintenance tasks ranking and selection for being jointly scheduled with production orders was put forward. The proposed approach was implemented based on a two-stage assessment method, which makes use of the DMCDM. The DMCDM enables to merge and jointly process and analyse maintenance information regarding historical, current, and provisional data, based on corresponding subsets of criteria, which are defined according to a group of decision- makers that interact on its definition and application of the proposed underlying maintenance tasks processing methodology, which is accessible through a developed CMS, accessible by a set of entities for enabling joint decision-making. The utilisation of the proposed GDM approach was illustrated through an industrial example of application and it revealed to be promising in supporting joint maintenance and manufacturing orders processing, once permitting to rank and select a set of maintenance tasks with the highest scores for being jointly scheduled with production orders by using other functionalities included in the CMS. This is a novel contribution, as far as our knowledge, and based on the study conducted there are no similar contributions in the literature that enable a distributed and dynamic maintenance tasks assessment and selection, based on a DMCDM, for being further jointly programmed with production orders, through the CMS. Besides, the CMS includes another functionality, namely, for predicting maintenance key performance indicators, which are considered through criteria included in the prognostic data processed using the DMCDM, such as mean time before failure. Thus, this work contributes to the maintenance and production orders management scientific domain, which continues lacking contributions that enables CDM, which is considered of utmost importance to promote a sustainable development of companies, and is supported by new technologies underlying the current digital age, being still necessary for further developments and industrial applications to be explored.

Acknowledgement

This work was supported by national funds through the FCT - Fundação para a Ciência e Tecnologia through the R&D Units Project Scopes: UIDB/00319/2020, and EXPL/EME-SIS/1224/2021.

References

Aggarwal, K., Atan, O., Farahat, A., Zhang, C., Ristovski, K., and Gupta, C. (2018). Two birds with one network: Unifying failure event prediction and time-to-failure modelling.

Arrais-Castro, A., Varela, M. L. R., Putnik, G., Ribeiro, R., and Dargam, F. (2015a). Collaborative negotiation platform using a dynamic multi-criteria decision model. *International Journal of Decision Support System Technology*, *7*(1), 1–14. IGI Global, USA. DOI: 10.4018/ijdsst.2015010101.

Arrais-Castro, A., Varela, M. L. R., and Putnik, G. D. (2015b), Spatial-temporal business partnership selection in uncertain environments., *FME Transactions*, *43*(4), 353–361. Belgrade University. DOI: 10.5937/fmet1504353A.

Arrais-Castro, A., Varela, M. L. R., Putnik, G. D., Ribeiro, R. A., Machado, J., and Ferreira, L. (2018). Collaborative framework for virtual organisation synthesis based on a dynamic multi-criteria decision model. *International Journal of Computer Integrated Manufacturing*, 1–12. Taylor & Francis. DOI: 10.1080/0951192X.2018.1447146.

Assaf, D. P., Scarf, P. R., and Iung, B. (2019). Modelling and application of condition-based maintenance for a two-component system with stochastic and economic dependencies. *Reliability Engineering & System Safety, 182*, 86–97.

Aydemir, G., and Acar, B. (2020). Anomaly monitoring improves remaining useful life estimation of industrial machinery. *J. Manuf Syst.*, 56, 463–469. https://doi.org/10.1016/j.jmsy.2020.06.014.

Bajestani, M. A., Banjevic, D., and Beck, J. C. (2014). Integrated maintenance planning and production scheduling with Markovian deteriorating machine conditions. International Journal of Production Research, *52*(24), 7377–7400.

Bakshi, T., Sinharay, A., Sarkar, B., and Sanyal, S. K. (2015). A New DST-Belief Theoretic Project Selection Model for Multi-Criteria Decision Support System. *Journal of The Institution of Engineers (India): Series C, 96*(4), 337–349.

Bektas, O., Marshall, J., and Jones, J. A. (2020). Comparison of computational prognostic methods for complex systems under dynamic regimes: A review of perspectives. *Archives of Computational Methods in Engineering*, 27, 999–1011.

Benker, M., Furtner, L., Semm, T., and Zaeh, M. F. (2021). Utili sing uncertainty information in remaining useful life estimation via Bayesian neural networks and Hamiltonian Monte Carlo. *Journal of Manufacturing Systems, 61*, 799–807.

Bevilacqua, M., and Braglia, M. (2000). The analytic hierarchy process applied to maintenance strategy selection, *Reliability Engineering & System Safety, 70*(1), 71–83.

Bhatt, G. D. and Zaveri, J. (2002). The enabling role of decision support systems in organisational learning. *Decision Support Systems, 32*(3), 297–309.

Biondi, M., Sand, G., and Harjunkoski, I. (2017). *Optimization of Multipurpose Process Plant Operations: A multi-time-scale maintenance and production scheduling approach*, Elsevier, Computers and Chemical Engineering.

Brannback, M. (1994). Decision support systems for strategic management. *Journal of Decision Systems, 3*(2), 95–114.

Campanella, G., Pereira, A., Ribeiro, R. A., and Varela, M. L. R. (2012). Collaborative Dynamic Decision Making: A Case Study from B2B Supplier Selection. *In: Decision Support Systems –Collaborative Models and Approaches in Real Environments*. Hernández, J.E., Zarate, P., Dargam, F. Delibašic, B., Liu, S. and Ribeiro, R. (Eds.), Lecture Notes in Business Information Processing (LNBIP), Vol. 121, 88–102. Springer Berlin Heidelberg. DOI: 10.1007/978-3-642-32191-7_7.

Chen, C. T. (2000). Extensions of the TOPSIS for group decision-making under a fuzzy environment. *Fuzzy Sets & Systems, 114*(1), 1–9.

Chen, C. T., Lin, C. T., and Huang, S. F. (2006). A fuzzy approach for supplier evaluation and selection in supply chain management. *International Journal of Production Economics, 102*(2), 289–301.

Costa J. P., Melo P., Godinho P., and Dias L. C. (2003). The AGAP system: A GDSS for project analysis and evaluation, *European Journal of Operational Research, 145*(2), 356–367.

Fitouhi, M. C., Nourelfath, M., and Gershwin, S. B. (2017). Performance evaluation of a two machine line with a finite buffer and condition-based maintenance. *Reliability Engineering & System Safety, 166*(March), 61–72.

Ghaleb, M., Taghipour, S., Sharifi, M., and Zolfagharinia, H. (2020). Integrated production and maintenance scheduling for a single degrading machine with deterioration-based failures. *Computers & Industrial Engineering, 143*, 106432.

Gong, X. and Qiao, W. (2014). Current-based mechanical fault detection for direct-drive wind turbines via synchronous sampling and impulse detection, *IEEE Transactions on Industrial Electronics, 62*(3), 1693–1702.

Jaramillo, P., Smith, R. A., and Andreu, J. (2005). Multi-Decision-Makers Equalis er: A Multi-objective Decision Support System for Multiple Decision-Makers. *Annals of Operations Research, 138*, 97–111.

Jassbi, J. J., Ribeiro, R. A., and Varela, M. L. R. (2014). Dynamic MCDM with future knowledge for supplier selection. *Journal of Decision Systems, Knowledge-Based Decision Systems, 23*(3), 232–248,

Jia, X., Huang, B., Feng, J., Cai, H., and Lee, J. (2018). Review of PHM data competitions from 2008 to 2017: methodologies and analytics. *Proc. Annual Conf. PHM Soc., 10*(1).

Keenan, P. (2016). Changes in DSS disciplines in the Web of Science. *Journal of Decision Systems, 25*(S1), 542–549.

Kim, N. H., An, D., and Choi, J. H. (2017). *Prognostics and Health Management of Engineering Systems*. Switzerland: Springer International Publishing.

Knoben, J., and Oerlemans, L. A. (2006). Proximity and inter-organis ational collaboration: A literature review. *International Journal of Management Reviews, 8*(2), 71–89.

Ladj, A., Varnier, C., and Tayeb, FB-S. (2016). IPro-GA: An integrated prognostic based GA for scheduling jobs and predictive maintenance in a single multifunctional machine. *IFAC-PapersOnLine 2016, 49*(12): 1821–1826. https://doi.org/10.1016/j. ifacol.2016.07.847.

Lee, C. Y., and Chen, Z. L. (2000). Scheduling jobs and maintenance activities on parallel machines. *Naval Research Logistics (NRL), 47*(2), 145–165.

Lee, K. W., and Huh, S. Y. (2006). A model-solver integration framework for autonomous and intelligent model solutions. *Decision Support Systems, 42*(2), 926–944.

Leyva Lopez, J. C., Alvarez Carrillo, P. A., Gastelum Chavira, D. A., and Solano Noriega, J. J. (2016). A web-based group decision support system for multicriteria ranking problems. *Operational Research*. Doi:10.1007/s12351-016-0234-0.

Li, N., Gebraeel, N., Lei, Y., Bian, L., and Si, X. (2019). Remaining useful life prediction of machinery under time-varying operating conditions based on a two-factor state-space model. *Reliability Engineering & System Safety, 186*, 88–100. https://doi.org/10.1016/j. ress.2019.02.017.

Limayem, M., Banerjee, P., and Ma, L. (2006). Impact of GDSS: Opening the black box. *Decision Support Systems, 42*(2), 945–957.

Liu, S., Duffy, A. H. B., Whitfield, R. I., and Boyle, I. M. (2010). Integration of decision support systems to improve decision support performance. *Knowledge and Information Systems, 22*(3), 261–286.

Luo, B., Wang, H., Liu, H., Li, B., and Peng, F. (2018). Early fault detection of machine tools based on deep learning and dynamic identification. *IEEE Transactions on Industrial Electronics, 66*(1), 509–518.https://doi.org/10.1109/TIE.2018.2807414.

Malhotra, P., Tv, V., Ramakrishnan, A., Anand, G., Vig, L., Agarwal, P., and Shroff, G. (2016). *Multi-sensor Prognostics using an Unsupervised Health Index Based on LSTM Encoder-decoder*. arXiv preprint arXiv:1608.06154.

McCall, J. J. (1956). Maintenance policies for stochastically failing equipment: A survey. *Management Science, 11*, 493–524.

Michau, G., and Fink, O. (2019, June). Unsupervised fault detection in varying operating conditions. *In: 2019 IEEE International Conference on Prognostics and Health Management (ICPHM)*, pp. 1–10 . IEEE.

Montero Jimenez, J. J., Schwartz, S., Vingerhoeds, R., Grabot, B., and Salaün, M. (2020). Towards multi-model approaches to predictive maintenance: A systematic literature survey on diagnostics and prognostics. *J. Manuf. Syst., 56*, 539–557. https://doi.org/ 10.1016/j.jmsy.2020.07.008.

Mosheiov, G., and Sarig, A. (2009). Scheduling a maintenance activity and due-window assignment on a single machine. *Computers & Operations Research, 36*(9), 2541–2545.

NAMUR (2009). *NAMUR-Recommendation NE129: Plant Asset Management*. International user association of automation technology in process industries (NAMUR).

Pan, E., Liao, W., and Xi, L. (2012). A joint model of production scheduling and predictive maintenance for minimis ing job tardiness. *International Journal of Advanced Manufacturing Technology, 60*(9–12), 1049–1061.

Putnik, G. D., Pabba, S. K., Manupati, V. K., Varela, M. L. R., and Ferreira, F. (2021), semi-Double-loop machine learning based CPS approach for predictive maintenance in manufacturing systems based on machine status indications. *CIRP Annals - Manufacturing Technology, 70*(1), 365–368. Elsevier, Netherlands. ISSN 0007-8506. https://doi.org/10.1016/j.cirp.2021.04.046

Qureshi, M. R., Singh, R. K., and Hasan, M. A. (2017). Decision support model to select crop pattern for sustainable agricultural practices using fuzzy MCDM. *Environment, Development and Sustainability*, DOI: 10.1007/s10668-016-9903-7.

Rahmati, S. H. A., Ahmadi, A., and Govindan, K. (2018). A novel integrated condition-based maintenance and stochastic flexible job shop scheduling problem: Simulation-based optimization approach. *Annals of Operations Research, 269*, 583–621.

Ran, Y., Zhou, X., Lin, P., Wen, Y., and Deng, R. (2019). *A Survey of Predictive Maintenance: Systems, Purposes, and Approaches*. arXiv preprint arXiv:1912.07383.

Razmak, J. and Aouni, B. (2015). Decision Support System and Multi-Criteria Decision Aid: A State of the Art and Perspectives. *Journal of Multi-Criteria Decision Analysis, 22*(1–2), 101–117.

Sikorska, J. Z., Hodkiewicz, M., and Ma, L. (2011).Prognostic modelling options for remaining useful life estimation by industry. *Mech. Syst. Signal Process, 25*(5): 1803–1836. https://doi.org/10.1016/j.ymssp.2010.11.018.

Simonov, K.-S., Varela, M. L. R., Putnik, G. D., Agyemang, J. B., and Ávila, P. (2018). Supplier Evaluation and Selection: A Fuzzy Novel Multi-Criteria Group Decision-Making Approach. *International Journal for Quality Research, 12*(2). University Montenegro.

Sloan, T. W., and Shanthikumar, J. G. (2000). Combined production and maintenance scheduling for a multiple-product, single-machine production system. *Production and Operations Management, 9*(4), 379–399.

Song, Y., Shi, G., Chen, L., Huang, X., and Xia, T. (2018). Remaining useful life prediction of turbofan engine using hybrid model based on autoencoder and bidirectional long short-term memory. *J. Shanghai Jiaotong Univ. (Sci.), 23*(S1), 85–94. https://doi.org/10.1007/s12204-018-2027-5.

Sprague, R. H., and Carlson, E. D. (1982). *Building Effective Decision Support Systems*. Prentice-Hall, Englewood Cliffs, NJ.

Staufen. (2018). *Deutscher Industrie 4.0 Index 2018*.

Taghipour, S., and Azimpoor, S. (2018). Joint optimization of jobs sequence and inspection policy for a single system with two-stage failure process. *IEEE Transactions on Reliability, 67*(1), 156–169.

Taha, Z., and Rostam, S. (2012). A hybrid fuzzy AHP-PROMETHEE decision support system for machine tool selection in a flexible manufacturing cell. *Journal of Intelligent Manufacturing, 23*(6), 2137–2149.

Taylor & Francis. DOI: 10.1080/12460125.2014.886850. http://www.scimagojr.com/journalsearch.php?q=19700181431&tip=sid&clean=0. (https://www.tandfonline.com/doi/full/10.1080/12460125.2014.886850).

Vafaei, N., Ribeiro, R. A., Camarinha-Matos, L. M., and Valera, L. R. (2019). Normali sation techniques for collaborative networks. Kybernetes, *49*(4), 1285–1304.

Varela, M. L. R., and Ribeiro, R. A. (2003). Evaluation of Simulated Annealing to solve fuzzy optimization problems. *Journal of Intelligent and Fuzzy Systems, 14*(2), 59–71, IOS Press. (https://content.iospress.com/articles/journal-of-intelligent-and-fuzzy-systems/ifs00204).

Varela, M. L. R., Arrais-Castro, A., and Ribeiro, R. A. (2018). A data fusion approach for business partners selection. *International Journal of Information and Decision Sciences, 10*(4), 311–344, Inderscience.

Varela, M. L. R., Putnik, G. D., Manupati, V. K., Rajyalakshmi, G., Trojanowska, J., and Machado, J. (2021). Integrated process planning and scheduling in networked manufacturing systems for I4.0: A review and framework proposal. *Wireless Networks, 27*(3), 1587–1599. DOI: 10.1007/s11276-019-02082-8.

Varela, M. L. R., Putnik, G. D., and Romero, F. (2022). The Concept of Collaborative Engineering: A Systematic Literature Review. *Production & Manufacturing Research, 10*(1), 784–839, Taylor & Francis. https://doi.org/10.1080/21693277.2022.2133856.

Varela, M. L. R., and Putnik, G. D. (Editors) (2022). Collaborative and Intelligent Networks and Decision Systems and Services for Supporting Engineering and Production Management (Editorial). *Future Internet,* MDPI.

Varela, M. L. R., Ávila, P., Castro, H., Putnik, G. D., Fonseca, L. M. C., and Ferreira, L. (Editors) (2022a). Manufacturing and Management Paradigms, Methods and Tools for Sustainable Industry 4.0-Oriented Manufacturing Systems (Editorial). *Sustainability, 14,* 1574 (2022), MDPI. https://doi.org/10.3390/su14031574.

Varela, M. L. R., Alves, C. F. V., Santos, A. S., Vieira, G. G., Lopes, N., and Putnik, G. D. (2022b). Analysis of a Collaborative Scheduling Model Applied in a Job Shop Manufacturing Environment (2022b). *Machines, MDPI, 10*(12), 1138, 1–16, https://doi.org/10.3390/machines10121138.

Varela, M. L. R., Putnik. G. D., and Romero, F. (2023). Collaborative Manufacturing and Management contextualization in the Industry 4.0 based on a Systematic Literature Review. *International Journal of Management Science and Engineering Management,* Taylor & Francis. https://doi.org/10.1080/17509653.2023.2174200.

Varela, M. L. R., Trojanowska, J., Cruz-Cunha, M. M., Pereira, M. A., Putnik, G. D., and Machado, J. (2023). Global Resources Management: A Systematic Review and Framework Proposal for Collaborative Management of CPPS. *Applied Sciences, 13*(750), 1–19, MDPI, Switzerland. DOI: 10.3390/app13020750.

Vieira, G. G., Varela, M. L. R., Putnik, G. D., and Machado, J. (2018). Intelligent Platform for Supervision and Production Activity Control in Real Time. *In: Advances in Manufacturing,* 151–159. Springer. DOI: 10.1007/978-3-319-68619-6_15.

Wang, Y. C., and Usher, J. M. (2005). Application of reinforcement learning for agent-based production scheduling. *Engineering Applications of Artificial Intelligence, 18*(1), 73–82.

Wang, T., Yu, J., Siegel, D., and Lee, J. (2008). *A Similarity-based Prognostics Approach for Remaining Useful Life Estimation of Engineered Systems.* pp. 1–6. https://doi. org/10.1109/PHM.2008.4711421.

Wang, S., and Yu, J. (2010). An effective heuristic for flexible job-shop scheduling problems with maintenance activities. *Computers & Industrial Engineering, 59*(3), 436–447.

Wang, P., Ananya Yan, R., and Gao, R. X. (2017). Virtualization and deep recognition for system fault classification. *J. Manuf. Syst., 44,* 310–316. https://doi.org/10.1016/j.jmsy.2017.04.012.

Wang, J., Zhang, L., Duan, L., and Gao, R. X. (2017). A new paradigm of cloud based predictive maintenance for intelligent manufacturing. *Journal of Intelligent Manufacturing, 28*(5), 1125–1137.

Wang, H., Yan, Q., and Zhang, S. (2021). Integrated scheduling and flexible maintenance in deteriorating multi-state single machine systems using a reinforcement learning approach. *Advanced Engineering Informatics, 49,* 101339.

Yang, S. L., Ma, Y., Xu, D. L., and Yang, J. B. (2011). Minimising total completion time on a single machine with a flexible maintenance activity. *Computers & Operations Research, 38*(4), 755–770.

Yoon, A. S., Lee, T., Lim, Y., Jung, D., Kang, P., Kim, D. et al. (2017). *Semi-supervised Learning with Deep Generative Models for Asset Failure Prediction.*

Zandieh, M., Khatami, A. R., and Rahmati, S. H. A. (2017). Flexible job shop scheduling under condition-based maintenance: Improved version of imperialist competitive algorithm. *Applied Soft Computing, 58,* 449–464.

Zarate, P. (1991). The process of designing a DSS: A case study in planning management. *European Journal of Operational Research, 55,* 394–402.

Zhai, S., Achatz, S., Groher, M., Permadi, J., and Reinhart, G. (2020). An empirical expert study on the status quo and potential of predictive maintenance in industry. *Int. Conf. Sens. Diagn. Progn. Control.*

Zhang, J., Wang, P., Yan, R., and Gao, R. X. (2018). Long short-term memory for machine remaining life prediction. *J. Manuf. Syst., 2018,* (48), 78–86. https://doi.org/10.1016/j.jmsy.2018.05.011.

Zhang, K., Chen, J., Zhang, T., He, S., and Pan, T. (2020). Zhou, Z. Intelligent fault diagnosis of mechanical equipment under varying working conditions via iterative matching network augmented with selective Signal reuse strategy. *J. Manuf. Syst.*, 2020(57), 400–415. https://doi.org/10.1016/j.jmsy.2020.10.007.

Zhai, S., Gehring, B., and Reinhart, G. (2021). Enabling predictive maintenance integrated production scheduling by operation-specific health prognostics with generative deep learning. *Journal of Manufacturing Systems, 61,* 830–855.

6

Blockchain-based Multi-agent System Framework for Collaborative Distributed Manufacturing System

In recent years, new collaborative manufacturing models such as distributed manufacturing and social manufacturing have attracted much attention as they have the potential to further transform existing production models and industrial structures. Blockchain as a new distributed computing architecture is successfully used in manufacturing to solve problems such as interoperability and collaboration, security and surveillance, marketing and protocols, democratic organization, global value chain management, etc. However, despite the utilization of agent-based technology to negotiate contracts among peers, security concerns persist. To address these challenges, this chapter proposes a blockchain-based multi-agent distributed framework tailored for the manufacturing sector. that aims to provide new ideas for co-design, implementation, and optimization of production and facilitate the upgrading and transformation of manufacturing industries.

1. Introduction

To minimize expenditures and enhance productivity, conventional manufacturing frequently depends on centralized, standardized large-scale production and customized mass production. However, this approach presents challenges. Static company configurations make it difficult to change, expand and integrate. The constrained product paradigm struggles to cater to the demands of individual users, and the system's rigidity is unsuitable for the dynamic oversight of production tasks.

With the rise of the worldwide marketplace and the onset of the digital economic era, product markets and consumer demands have grown progressively fragmented. Labour and collaboration networks within social spheres are improving, and personalized, consumer-centric product innovation is emerging as the latest trend, aided by advanced technologies like 3D printing and RFID. Lean, modular, cooperative production approaches such as distributed manufacturing (DM) and social manufacturing (SM) are now feasible.

While the potential of these cooperative manufacturing methods is bright, they encounter diverse needs spanning technology, services, infrastructure, and resources, particularly given the gradual refinement of associated elements and institutional components. Blockchain technology is anticipated to serve as the optimal digital foundation for integration. Furthermore, when paired with blockchain-powered smart contracts, programmable assets enable the establishment of secure and streamlined information exchange, value transfer, and asset administration, ultimately leading to a significant overhaul of business models and existing social production relations.

Classical manufacturing scheduling operates under the assumption that manufacturing resources are situated in close geographic proximity. However, in reality, these resources, known as distributed manufacturing resources, are dispersed across various enterprises located in disparate spatial locations. Addressing the challenge of coordinating shared scheduling of distributed manufacturing resources (SSDMR) proves complex and arduous. Multi-agent technology assumes a crucial role in managing intricate, dynamic, and decentralized scheduling dilemmas. To tackle the present SSDMR predicament, two distinct architectures of multi-agent systems are devised. One constitutes an enterprise multi-agent subsystem, featuring a hierarchical arrangement, abstracted from the conventional hierarchical structure of manufacturing enterprises. The other entails an enterprise alliance multi-agent system, characterized by a federated architecture. Subsequent sections expound upon the SSDMR issue and elucidate the shared architecture of multi-agent systems.

As DM, Multi agent Systems, Blockchain are relatively new concepts that have yet to be fully realized. Many scholars have proposed conceptual models and frameworks to guide and accelerate their creation but the potential impact of blockchain has not been adequately captured. As a first attempt in this area, building on existing work, this chapter initially classifies the challenge of achieving the aforementioned mode of production into five several groups: interoperability and collaboration, security and monitoring, marketing and protocols, democratic organization, global governance and value chains. Taking into account these challenges and corresponding blockchain solutions, as well as key processes, the blockchain based multi agent distributed manufacturing system has been developed.

The article follows this structure: Section 2 surveys the literature and introduces the basic concepts; Section 3 describes the challenges; Section 4 presents a blockchain-based framework for collaborative manufacturing and demonstrates it through practical application examples. Finally, Section 5 presents the conclusion.

2. Literature Review

2.1 Block chain

The technological foundation of Bitcoin, Blockchain, comprises a sequence of linked data blocks forming a decentralized ledger, which is collectively maintained and shared by every node within the system. In this blockchain ecosystem, nodes, referred to as miners, establish connections and engage in peer-to-peer communication networks incentivized by mechanisms driving their participation.

Auto computing power can be provided for verification and packaging spread. during a certain period of time. At the same time, miners compete for settlement privileges based on a consensus mechanism in which the winner connects its assigned block to the main chain and receives a corresponding reward, which is subsequently updated by other nodes.

The term "smart contract" was initially introduced in 1994 by computer scientist and cryptographer (Matt et al., 2015)). Szabo (2019) defined it as "a collection of digitally encoded commitments, accompanied by protocols that govern the fulfil ment of those commitments", incorporating regulations and conditional responses. These contracts have the capability to encompass, authenticate, and carry out intricate transactions among distributed nodes, facilitating the exchange of information, value, and management of assets. As transactional protocols executed by computers capable of autonomously validating and enforcing terms and conditions sans intermediaries, intelligent digital contracts possess the ability to merge with diverse assets, transactions, and data, functioning as reliable agents to execute contracts securely and efficiently. This capability paves the way for a broad spectrum of intelligent assets and systems, establishing a distinct category within the digital realm.

Distributed ledger technology (DLT) and smart contracts powered by DLT share fundamental traits including trustlessness, decentralization, autonomy, anonymity, traceability, and resistance to tampering. These attributes offer extensive potential applications and have garnered significant interest. Presently, they find utilization across diverse sectors such as healthcare, finance, and the internet.

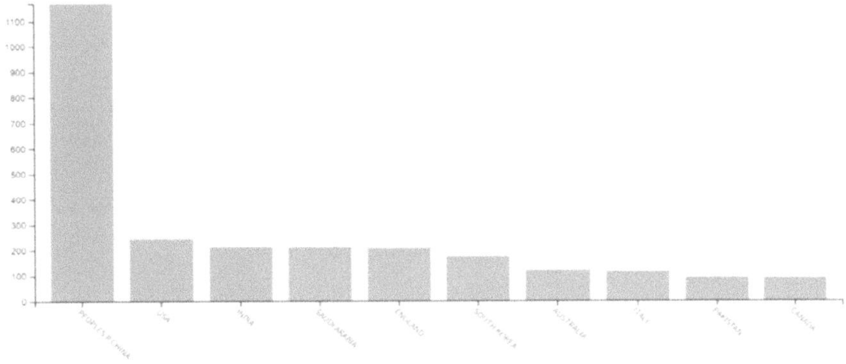

Figure 1. The research articles published country- wise (Top 10) in the areas of blockchain, multi-agent and distributed manufacturing systems in the last 10 years.

2.2 *Distributed Manufacturing*

A DM system represents a production framework that includes small manufacturing units and is supported by advances in physics, digitalization, and communication technologies (Jiang et al., 2016). This configuration locates production assets and enables instant communication within the supply chain, promoting customer-centric mass customization and system flexibility, strengthens adaptability, agility, and flexibility. When a DM system is guided by ecological requirements for redistributing manufacturing elements, such as location, scope, standards, cost, risks, and responsibility, it moves to redistributed manufacturing (R dM) because RdM is currently considered a branch of DM and the proposed collaborative model is universally applicable; therefore the difference between these two concepts will not be emphasized below and they will be collectively referred to as DM.

DM: By digitally integrating the entire production cycle and optimizing logistics, products can be manufactured on demand in virtual formats without geographical restrictions, taking advantage of local production resources and accessible manufacturing technology. Digital design and product sharing promotes data-driven open innovation while redistributing stakeholder roles. Customers are elevated to the role of co-creators of value, actively participating in the democratization of the value chain. This allows organizations and companies to achieve sustainable development in all economic, ecological, social and political dimensions.

Figure 2. The number of research articles published in the areas of blockchain, multi-agent, and distributed manufacturing systems in the last 10 years.

Given the limitations arising from current technological and resource constraints, several obstacles remain in the practical implementation of DM Shang et al. (2019) and Yuan and Wang (2016). This work acknowledges common challenges and potential blockchain solutions and conducts a comprehensive review of the existing literature. The challenges identified are grouped into five main areas: interoperability and collaboration, security and oversight, market integration and protocols, democratic governance structures and global value chain management. Each of these categories is then presented and analyzed individually.

(1) Interactivity and collaboration: This presents significant challenges, including limited access to individuals and small and medium-sized enterprises (SMEs), complexities in interactions among multiple agents; challenges in task planning, difficulties in scheduling resources are involved, potentially leading to monopolistic tendencies. In distributed production environments involving multiple agents with different capabilities, immediate and efficient interactive networks are critically needed to facilitate accurate communication of production information, rational task planning and optimal resource utilization while avoiding conflicts in social mining (SM). It underlines the importance of interaction and collaboration among stakeholders, along with efficient resource allocation and tracking in outsourcing and crowdsourcing scenarios. These dynamic management processes involve coordinating network participants, social resources, and production services, a task that often goes beyond the capabilities of traditional social media platforms [6]

(2) Security and monitoring: This presents significant challenges, including difficulties in identity authentication, secure storage and sharing of data, protection of intellectual property and other legal rights, monitoring and auditing challenges. Besides being difficult and facing threats from fake and malicious nodes, since product data transfer replaces physical transport of product, it is important to ensure secure storage and exchange of data containing critical information. It is important to create an enabling environment to engage in construction and initiatives. In addition, the implementation of robust certification systems can facilitate forensic analysis and monitoring of interactions among groups with multiple attributes and interests.

(3) Marketing and protocols: It presents significant barriers, including challenges in engaging individual users and production intermediaries, difficulties in reaching tailored agreements, and complexities of risk distribution. Valid conformity with production units relevant to individual consumers as well in an optimal digital manufacturing (DM) system. There should be an opportunity to enter production agreements, allowing for greater product customization. This highlights the need for improved engagement mechanisms for production intermediaries, facilitating tailored agreements and efficient contract negotiation forums. To close this gap, there is an urgent need to digitalize production intermediaries and optimize order agreements through more flexible, standardized, accurate, intelligent, and reliable approaches. In the context of social mining (SM), there are challenges to effectively match social real-time needs with production capabilities and lack proper safety mechanisms.

3. Proposed Framework for Block chain-based Agent for Transparent Negotiation between Various Entities in DMS

Multi-agent systems (MAS) belong to the field of distributed artificial intelligence (DAI). They are interconnected autonomous entities that work together in a collaborative environment to achieve specific goals Matt et al., 2015. MAS is widely used in fields such as finance, energy, and electronic health (eHealth) and is recognized for its adaptability, cost-effectiveness, and efficiency Wang et al., 2019; Petrulaityte et al., 2017; Stewart and Tooze, 2015; Srai et al., 2016) solve the expected complex problems; however, vulnerabilities in system security, transparency, and coordination have been identified, which threaten the integrity of the system (Rauch et al., 2016) .

4. Proposed Methodology of Blockchain

Recent studies have shown that integrating blockchain technology (BCT) is a viable solution Economist; Chung et al., 2019; Deng and Zhang, 2019). Although this proposition fits well with the characteristics of BCT and its applicability in various fields, including those overlapping with MAS Burke et al., 2017, the existing literature highlights the rigour and implementation gap of this integration. BCT is based on distributed computing and has a distributed accounting system that works on a peer-to-peer basis. BCT is often associated with Bitcoin and other cryptocurrencies, but has applications in a variety of fields. The unique hash encryption ensures high security and the decentralized nature minimizes risks. In addition, consensus protocols facilitate coordinated control and transparent recording of transactions on the blockchain promotes trust by maintaining an immutable ledger. Integrating BCT into MAS not only improves security but also provides system-wide optimizations to improve performance.

Technically speaking, the MAS framework provides a viable structure for integration with BCT. Among them, JADE (Java Agent Development Framework) stands out as the most popular and widely used platform for MAS. In contrast, the Ethereum network serves as the framework platform for BCT.

A conceptual framework for integrating the BCT and MAS frameworks is presented below and comes from Fornasiero and Carpanzano (2017). In this architecture, the CA-A1 BCT represents the CA agent, which is responsible for providing agent functionality to interact with the network. BC-A represents a typical MAS agent, having the appropriate capabilities and requiring certification (eCert) by a CA agent.

The architecture shown in Figure 3 operates between different agents, such as a customer agent and a corporate agent, and facilitates coordination between customer requests and corporate agent responses. The information is then verified by the plant agent before being sent to the producer agent within the DMS. The transaction or negotiation is successful if the response from the producer agent is received and accepted. Otherwise it is unsigned. Similarly, requests are sent to all other production facilities according to priority.

……

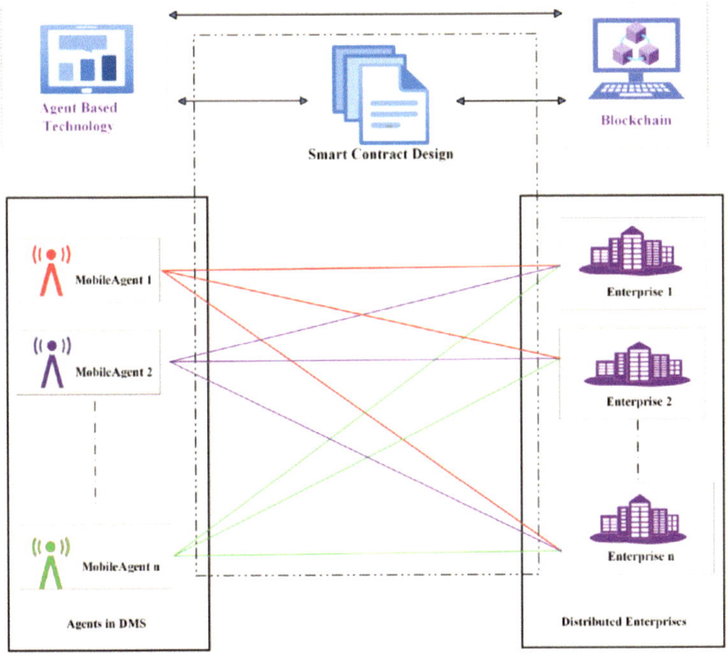

Figure 3. Block-chain based smart contract for mobile agent technology in DMS.

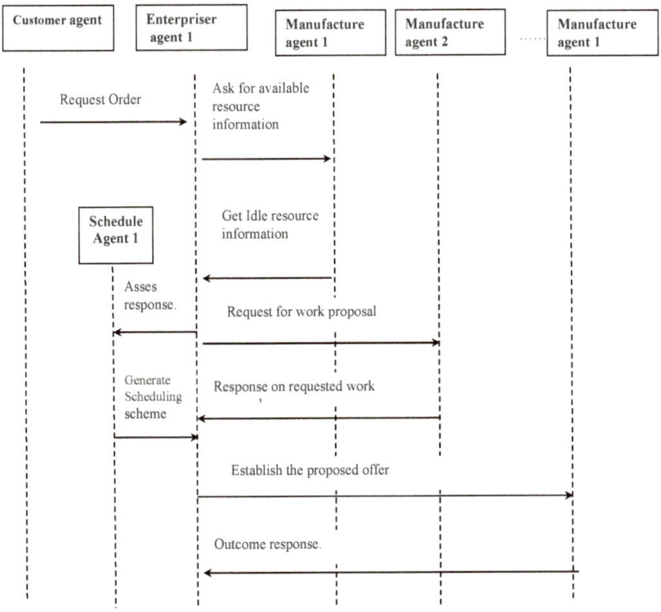

Figure 4. Proposed scheme for smart contract between customer agent, enterprise agent, and manufacturer agent.

Table 1. Pseudo code for the smart contract between customer agent and enterprise agent.

Algorithm 1: Sharing Request between Client Agent and Enterprise Agent for Product Request

Input: product_name, product_quantity, product_color, expected_delivery_date

7. Initialize Integer Order Sequence to 0 and a Mapping from integer to order structure called orders.
8. O ← (product_name, product_quantity, product_color, expected_delivery_date)
9. Create the order structure O and store it.
10. Orders [Order Sequence++] ← O (storing the order)

Algorithm 2: Query Product Request Input: Order_id (Integer) Output: Order Object

1. O ← orders[Order_id]
2. Return O

Algorithm 3:

Define the contract " Enterprise agent 1 and Manufacturing agent

Declare two public integer variables to store the engine quality and the required quality

Function to set enterprise agent request

Function to set manufacturer agent response

Function to check the deal

function checkDeal()

If the Requested service is available the Response positive

if (Required request from agent;= available capacity)

return "Deal successful"

Otherwise, return "Deal unsuccessful"

5. Results of the Proposed Proof of Concept in the Considered DMS

In this scenario, we consider a DMS. Each manufacturing entity has the ability to produce products according to different customer orders. In this scenario, customer agents are initially negotiating a deal with enterprise agents. Corporate users take responsibility for fulfilling orders requested by various customers, acting as brokers or intermediaries in the virtual corporate space, and maintaining the relationship between customers and the company. Each product request must be fulfil ed by different manufacturing services from different companies. The concept of our model is to implement the smart contract between various peers in the DMS with the help of Agent-based systems. Request matching process as a smart contract and execute this contract on a distributed blockchain-enabled platform across multiple nodes to execute the contract and store the results. The customer initiates an order, which is then accepted by the business user and passed on to the company to assess their ability to fulfil the customer agent request.

The proposed blockchain model (BC) consists of two smart contracts. The original agreement listed in Table 1 sets forth the agreement between the customer representative and the company representative. This unique smart contract simplifies the process of publishing requests, ensuring that requests are delivered efficiently and stored securely for later processing.

Subsequently, the second contract shown in Table 5 governs the interaction between the corporate agent and the producer agents. This agreement serves as a means of matching published requests with appropriate production agents who can provide the required services. Furthermore, Figure 4 shows the proposed model for using smart contracts in blockchain and describes their roles and interactions.

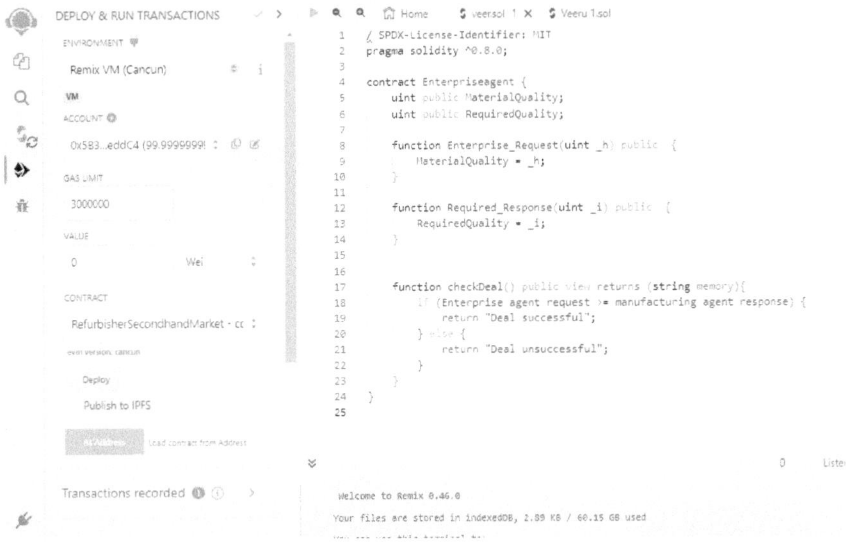

Figure 5. The proposed proof of concept for the proposed algorithm between the customer agent and enterprise agent.

A proof of concept was developed by a customer agent and an enterprise agent and is shown in Figures 5 and 6 in the screenshots above. The proposed system consists of the following elements implemented in an advanced configuration using intel. (R) Core(TM) i9-10900K 3.70GHz processor, running Ubuntu 20.04 LTS 64GB RAM. This implementation uses Remix Ethereum with Solidity Pragma version ^0.8.4 to implement the proposed system.

The smart contracts between the customer agent and enterprise agent have been developed and implemented on the Ethereum platform. The

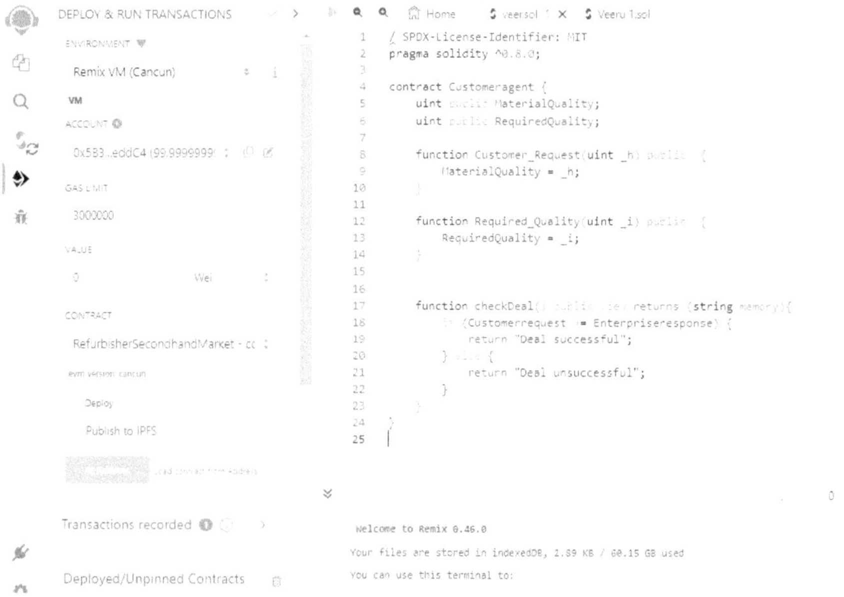

Figure 6. The proposed proof of concept for the proposed algorithm between the enterprise agent and manufacturer agent.

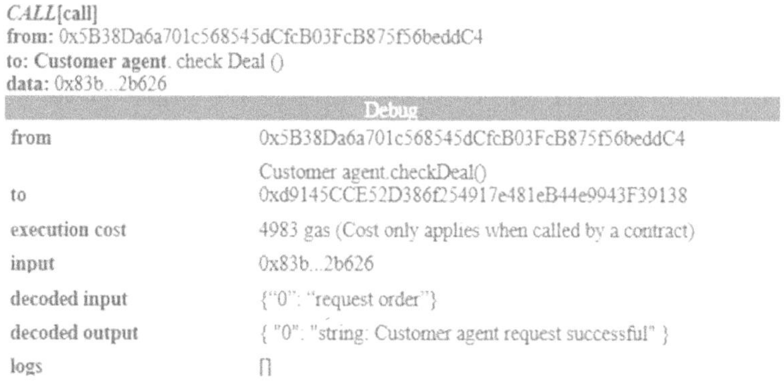
CALL[call]
from: 0x5B38Da6a701c568545dCfcB03FcB875f56beddC4
to: Customer agent. check Deal ()
data: 0x83b...2b626

Debug	
from	0x5B38Da6a701c568545dCfcB03FcB875f56beddC4
to	Customer agent.checkDeal() 0xd9145CCE52D386f254917e481eB44e9943F39138
execution cost	4983 gas (Cost only applies when called by a contract)
input	0x83b...2b626
decoded input	{"0": "request order"}
decoded output	{ "0": "string: Customer agent request successful" }
logs	[]

Figure 7. The screenshot for the smart contract between the enterprise agent and manufacturer agent.

transactions and results were recorded on the public permission less blockchain, as depicted in Figures 7 and 8.

The smart contracts between the enterprise agent and manufacturing agent have been developed and implemented on the Ethereum platform. The transactions and results were recorded on the public permission less blockchain, as depicted in Figure 9.

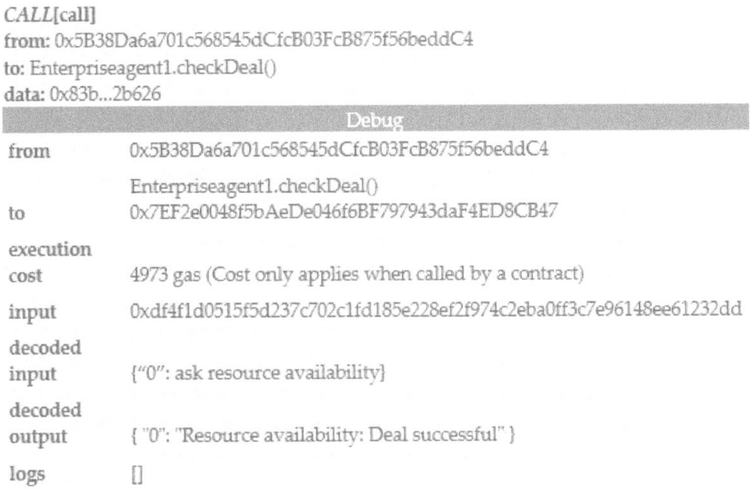

Figure 8. The screenshot for the smart contract between the enterprise agent and manufacturer agent.

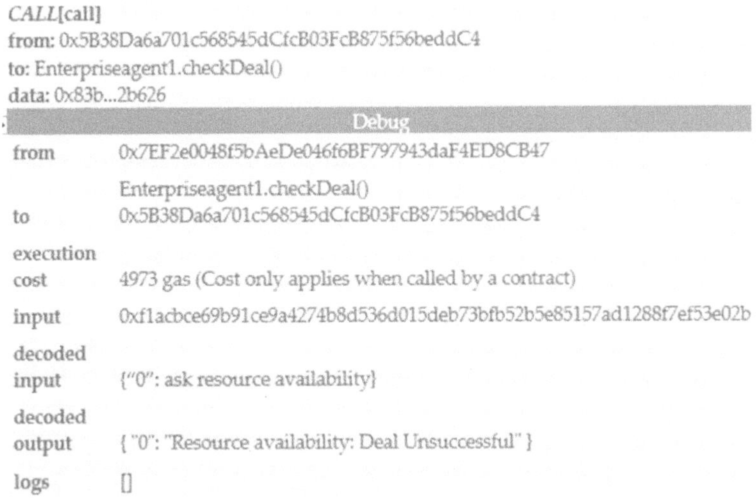

Figure 9. The screenshot for the smart contract between the enterprise agent and manufacturer agent.

6. Conclusions

In conclusion, current blockchain-based agent-based frameworks represent significant advances in the management of distributed manufacturing systems, and continuous efforts are underway to address scalability, security, and environmental considerations within this framework.

Research and development is essential. Because of their relevance, future work should focus on these aspects. The developed smart contracts where we increase the efficiency of the proposed framework and contribute, especially in contract negotiations between customers, enterprise s, and manufacturers in distributed manufacturing systems. In this scenario, numerous manufacturing enterprises scattered across diverse locations join forces to establish a Distributed Manufacturing System (DMS), with the goal of attaining competitive edges. One major obstacle encountered within the DMS is the requirement for manufacturing entities to depend unquestioningly on each other for conducting their activities. Such dependency limitations impede further exploration of the DMS within the fiercely competitive, consumer-oriented market. Consequently, businesses are in pursuit of advanced technological remedies to mitigate this trust issue. In this context, Agent-Driven Blockchain Technology offers several benefits, such as enhanced security and transparency, facilitating DMS entities to exchange resource information without relying blindly on one another. Prior studies have put forth various frameworks for Blockchain-based resource management within DMS. However, limited literature delves into the application of intelligent contracts in supply chain management to oversee and track products.

References

Burke, R., A. Mussomeli, S. Laaper, M. Hartigan, and B. Sniderman. The smart factory: Responsive, adaptive, connected manufacturing. Deloitte Insights, *31*(1), 1–10, 2017. https://www2.deloitte.com/content/dam/insights/us/articles/4051_The-s mart-factory/DUP_The-smart-factory.pdf.

Chung, K., H. Yoo, D. Choe, and H. Jung. Blockchain network based topic mining

Deng, J., and Q. Zhang. Approximating hypervolume and hypervolume contributions using polar coordinates. IEEE Trans. Evol. Comput., *23*(5), 913–918, 2019. https:// doi.org/10.1109/TEVC.2019.2895108.

Economist, A third industrial revolution, http://www.economist.com/ node/21552901, accessed July 15, 2019.

Fornasiero, R., and E. Carpanzano. Advances in customer-oriented manufacturing and value chain management. *30*(7), 677–679, 2017.

Jiang, P., J. Leng, and K. Ding. Social manufacturing: A survey of the state-of-the-art and future challenges, In: 2016 IEEE International Conference on Service Operations and Logistics, and Informatics (SOLI). IEEE, 2016, pp. 12–17.

Jiang, P., K. Ding, and J. Leng. Towards a cyber-physical-social-connected and service-oriented manufacturing paradigm: Social manufacturing, Manufacturing Letters, *7*, 15–21, 2016.

Matt, D. T., E. Rauch, and P. Dallasega. Trends towards distributed manufacturing systems and modern forms for their design, Procedia CIRP, *33*, 185–190, 2015.

Petrulaityte, A., F. Ceschin, E. Pei, and D. Harrison, Supporting sustainable product-service system implementation through distributed manufacturing, Procedia CIRP, *64*, 375–380, 2017.

process for cognitive manufacturing. Wireless Pers. Commun., *105*(2), 583–597, 2019. https://doi.org/10.1007/s11277-018-5979-8.

Rauch, E., P. Dallasega, and D. T. Matt. Sustainable production in emerging markets through distributed manufacturing systems (DMS), Journal of Cleaner Production, *135*, 127–138, 2016.

Shang, X., Z. Shen, G. Xiong, F.-Y. Wang, S. Liu, T. R. Nyberg, H. Wu, and C. Guo, Moving from mass customization to social manufacturing: a footwear industry case study, International Journal of Computer Integrated Manufacturing, *32*(2), 194–205, 2019.

Srai, J. S., M. Kumar, G. Graham, W. Phillips, J. Tooze, S. Ford, P. Beecher, B. Raj, M. Gregory, M. K. Tiwari, et al., Distributed manufacturing: Scope, challenges and opportunities. International Journal of Production Research, *54*(23), 6917–6935, 2016.

Stewart, H., Stewart and J. Tooze. Future makespaces and redistributed manufacturing. Making Futures, *4*, 1–9, 2015.

Szabo, N. Smart Contracts. http://www.fon.hum.uva.nl/rob/Courses/ InformationInSpeech/CDROM/Literature/LOTwinterschool2006/szabo.best.vwh.net/smart.contracts.html, accessed July 15, 2019.

Wang, S., L. Ouyang, Y. Yuan, X. Ni, X. Han, and F.-Y. Wang, Blockchain-enabled smart contracts: Architecture, applications, and future trends, IEEE Transactions on Systems, Man, and Cybernetics: Systems, 2019.

Yuan, Y., and F.-Y. Wang. Blockchain: The state of the art and future trends. Acta Automatica Sinica, *42*(4), 481–494, 2016.

7

Process Improvement in a Semi-automated Shoe Polish Manufacturing Company: A Simulation Study

In the manufacturing industry, process improvements are essential for effective operations planning. In this study, we attempted to mimic a shoe polish manufacturer's present operational approach in order to improve system performance by reducing bottlenecks. As a first step, the existing industry manufacturing processes have been thoroughly investigated to establish the techniques for future modifications. Later, with simulation analysis the As-is model has been validated to examine and analyse the performance measures such as makespan, throughput time, and work in progress. Then, utilising a range of realistic scenarios, we try to improve the system's performance metrics. The production manager reviews the results of the generated scenarios and approves for its future implementation.

1. Introduction

The global shoe care market was valued at $ 4249.4 million in 2019, and it is estimated to grow to $ 5224 million by 2026, with a 3.0% CAGR over the forecast period. The global shoe shiner market is predicted to grow as formal footwear sales rise due to strict rules requiring formal shoes. Expanding population, shifting fashion trends, rising online sales and disposable income, wearing different shoes for different situations and purposes, and changing lifestyles are all projected to contribute to the expansion of the shoe sales and shoe care market.

Every production system's primary goal is to produce high-quality products while making efficient use of available resources. Balancing these objectives partially contradicts the goals of upcoming Industry 4.0 concepts. Industry 4.0 aims to digitise the production process while decreasing human participation by embracing concepts such as digital twins (Zawadzki and Zywicki, 2016), virtual and augmented reality, and hybrid simulation. It envisions machine tools and production techniques becoming more compatible, flexible, and intelligent (Chen, 2019; Xu, 2017; ElMaraghy, 2021) to improve efficiency, productivity, and accessibility in traditional production systems (Brecher et al., 2017).

Several obstacles arose while restructuring the production system and machine tools for a sustainable and robust manufacturing setup. Modelling, simulation, and analysis to create a model that would better design the assembly line process of the product. To address the problem, we propose in this study the XYZ Contract Manufacturing plant and ABC Industries, both located in southern India and manufacturing shoe care goods, particularly shoe shiners, in a traditional setup that is labour demanding. Work study is important in labour-intensive industries such as footwear, footwear accessory manufacturing, garment, and agriculture because it improves task processing time, determines labour productivity, studies worker motion, material movement, and assembly- line efficiency. The goal of the task is to investigate existing procedures, standardise and balance the workload of all workstations on the selected assembly lines, and improve line efficiency. Also, offer relevant solutions to improve labour productivity through process automation, assembly line simulation, and layout perspective using a scientific methodology.

The remaining sections of this work are organised as follows. Section 2 provides a brief literature overview. Section 3 contains the case description. Section 4 explains the process. Section 5 explains the step-by-step experimenting approach. Section 6 contains the findings and conclusions, as well as remarks and potential future research areas.

2. Literature

Manufacturing is an important sector of many countries' GDP because it creates high-paying jobs, promotes technological innovation, and generates more economic activity than any other industry. Assembly takes up 40–60% of overall manufacturing time. To improve assembly scheduling, efforts are undertaken to match limited assembly resources with assembly jobs within a particular sequence and time constraints (Parente, 2020).

According to Guzman et al. (2022) and Yazdani et al. (2021), Assembly Sequence Planning (ASP) is also integral to production scheduling. Zhang et al. (2020) proposed an ASP technique based on an assembly precedence graph, using a basic firework algorithm to limit the number of changes in assembly direction and tool switching. While, Li et al. (2022) investigated the dynamic scheduling and assignment of due dates for assembly production where processing time is unknown and widely spread. Given the uncertainty of processing time and random machine failure, Zheng et al. (2022) developed a modified master-apprentice evolutionary algorithm to enhance the scheduling system's robustness. Liang et al. (2021) proposed a machine failure prediction approach based on a convolutional neural network (CNN) that triggers rescheduling when the machine fails.

Shahrabi et al. (2017) examined irregular work schedules and equipment malfunctions while deciding on a time node for rescheduling. Dual Q-learning, a reinforcement learning method, was used to overcome workshop scheduling issues, such as shifting task arrival rates. Luo et al. (2021) employed a Markov decision process to represent the dynamic flexible job shop problem, in which the agent selects the tasks to be handled next and the related machine, as well as learning the best scheduling rule at each rescheduling point.

Zhang et al. (2021) investigated disruption-induced rescheduling and developed a hybrid multi-population genetic algorithm (MPGA) and constraint programming (CP) MPGA-CP method for maximising makespan, maximum machine workload, and overall tardiness. Reinforcement learning, a unique technique, has been used in dynamic task scheduling due to its capacity to handle uncertainty in a dynamic setting, self-learning capabilities, processing efficiency, and adaptability. Johnson et al. (2022) created a Multi-Agent Reinforcement Learning system that schedules dynamically arriving assembly jobs in a robot assembly cell. In this study, we used simulation to design an assembly line for the company's shoe polish production process.

Simulation tools are utilised in a wide range of scientific and industrial applications. Computer simulations investigate the movements of various parts of human activity, such as motion study, to increase productivity and train specialised professionals such as pilots (Saastamoinen and Maunula,

2021), the study of more complex mechanics experiments (Fan, 2020), machine or production operators (Barosz et al., 2020), (Peruzzini, 2020), and high-voltage engineers (Wang et al., 2023). Advanced algorithms enable the creation of weather simulation models from data collected from multiple weather sensors (Jandaghian, 2020).

Analytical methods based on computer simulations are gaining popularity due to improved problem solution accuracy and considerable technological developments in computer science. Thus, the use of modelling and simulation in manufacturing engineering is unavoidable. They are used in manufacturing processes to inspect the complete process as well as to test the functionality of an object in a short period of time, such as an operation, activity, station, warehouse status, and so on. Modelling and simulation of manufacturing processes allows for the validation and implementation of guidelines prior to their application in the real model, as well as the detection of any errors that may develop during operation.

2.1 *Need for the Study*

Although the company is dedicated to maintaining high standards, it feels that a thorough analysis of plant productivity is necessary. This analysis should cover technology, labour, process, and layout. This study presents a simulation model for redefining assembly lines and semi-automated manufacturing processes using heavy manual operations. Increasing production and process efficiency is the purpose, and the goal is to develop an efficient operational process by identifying and eliminating the bottle necks. The model defines the assembly procedure for hybrid assembly lines, which combine manual labourers and automated assembly machinery, taking into account the production goals, product attributes, and assembly responsibilities. Ultimately, the efficacy of the simulation model was evaluated and used in an assembly process in the shoe polish sector.

3. Case Description

XYZ is a shoe polish factory located in southern India that employs 200 people with an 8-hour shift, twice a day. This labour-intensive company operates in a typical setting and carries out a range of manual and partially automated tasks. As seen in Figure 1, these tasks include polish filling, impregnation, foam drilling, cutting, pasting, labelling, and packing with conventional machines. Most of the company's sales are generated in Europe to the tune of about 80%, with the Asia-Pacific area accounting for the remaining 20%. Nonetheless, the tasks alternate between manual and automated manufacturing lines depending on demand.

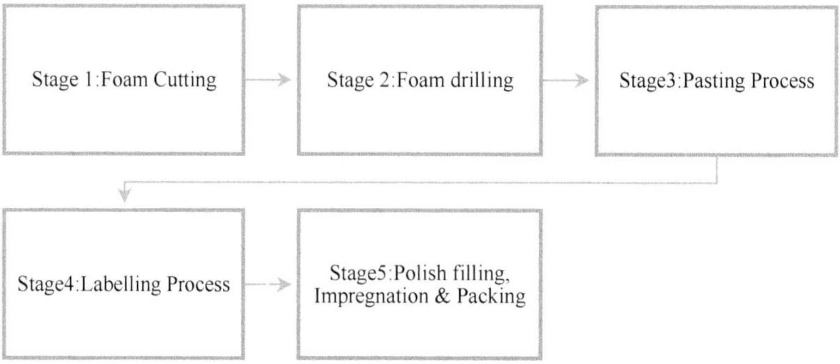

Figure 1. XYZ manufacturer activities.

The business is finding it difficult to keep up with the increasing demand from around the globe. Examining the current setup and protocols, comprehending the workload at each workstation for each assembly- line activity, and identifying the demand fulfilment constraint by using simulation tools to evaluate the system's throughput, cycle time, failure rate, and overall performance using scenario generation are all necessary to address the issue and offer solutions. In addition, it is imperative to suggest suitable measures to enhance worker productivity through process automation and assembly line simulation through a scientific methodology.

4. Methodology

The following steps were systematically used to conduct the simulation analysis for improving the system effectiveness.

I. Evaluate existing production processes, systems, procedures, and automation levels.

II. Conduct SOP analysis for current production processes and collaborate with management to identify obstacles and opportunities.

III. The study collected primary information on manufacturing/testing facilities, individual processes, automation levels, inventory (RM and FG), and more through discussions, questionnaires, and observations. The Plant Head coordinated and supported these discussions.

IV. Collect activity-based data from employees and supervisors in specific roles.

V. Simulation analyses were employed to examine manufacturing and support processes, as well as resource use.

5. Experimentation

The five processes of shoe polish manufacturing process are, beginning with foam cutting, followed by foam drilling and pasting, labelling, polish filling, impregnation, and packaging. The following steps are described in detail:

Stage 1: Foam Cutting Process

The foam cutting operation begins with the collection of foam sheets measured in length, width, and height. The foam used in the shoe shiner is an essential component of the manufacturing process. It is delivered in lots or bundles and then disassembled into pieces. To meet the market's strong demand for shoe shiners, daily average production is determined using the capacity of the production line, the maximum feasible output per day, per line, the number of hours per shift, and the worker's takt time. ABC presently has two identical production lines with semi-automated processes. So, in the event of a high demand, both production lines are activated to meet market demand.

ABC uses two hydraulic press machines grouped in a line to cut foam into the necessary shape, which are operated manually by two people, one for each machine. The foam sheets are kept near the machines so that they may be quickly picked up and placed on the cutting board. The worker must be skilled while positioning the die on the foam; otherwise, misplacement or inappropriate pressure wears and shreds the foam while also wasting foam and time. Before applying pressure to the hydraulic press machines, workers ensure that the die is properly positioned on the foam. One foam cutting takes 8 seconds.

i. Proper die placement on foam
ii. Applying appropriate pressure.
iii. Demolding: One person collects pressed foam sheets and meticulously removes each piece from the cut sheet.
iv. Determine the average time required for demoulding and quality control checks per sheet.
v. To address absenteeism, five more workers were educated to do quality control checks for air holes during demoulding, as this is a skilled task that is often done manually.

Stage 2: Foam Drilling

The foam from the previous station acts as an input for this station. Two professional workers are hired to manually put the foam pieces on the drill bit configuration in the automated drilling system. The foam drilling method produces fumes and heat; thus operators must wear hand gloves and respiratory masks. To drill the soft foam, hot wire coils (pugs) are used as drill bits. Operators press and hold the click switch simultaneously. During this procedure, the pug rotates up to one and a half times to drill the foam before the click switches are released. The drilled foam pieces are removed and placed in a separate tray. When the foams are removed, the drill bench is cleaned to eliminate waste from drilling. Brush cleaning is performed after 8 to 9 drilling operation cycles to remove foam particles. Each hour, the drill bits are removed from the drilling machine.

Stage 3. Pasting Process

The pasting apparatus consists of two hoopers set side by side, each carrying hot melt glue controlled by a single operator. One operator assembles the protection, while the other applies the foam to each hot melt glue gun. When the desired temperature is obtained, the operator applies hot glue to the sponge plate (average time: 3 to 5 seconds per plate). Within seconds after applying the glue, the foam pasting operator picks up the sponge plate, affixes a drilled foam to it, presses it evenly, and stacks it in a container. On an average, 16 pieces are pasted per nozzle every minute, while 64 pieces are pasted per minute with four nozzles.

Stage 4. Labelling Process

In the labelling unit, two automated machines work together with manual intervention to label the front and back. The required label, either front or back, is appropriately fed into the labelling machine. The sensor is inserted in position 1 or 2, depending on the material needed, cap or grip. An operator attaches the cap/grip on the pugs on the moving conveyor. The conveyor belt speed is programmed to circle at a specific rate per minute. The printing of batch codes (labels) depends on whether the product is being sent domestically or overseas. Batch coding is vertical in domestic markets, but horizontal in export markets. When demand is great, both labelling lines function well. Coding is done offline.

Stage 5: Polish Filling, Impregnation, and Palletization

The polish filling station is semi-automatic, requiring manual intervention at several places along the conveyor line. The conveyor speed is selected based on the process requirements. The polish filling operation starts with the silicone blend being filled to a predetermined level in the hopper. The mix level in the hopper is checked using the provided set-up and replenished when it reaches the cut-off level. The operator manually places the IBM container into the conveyor's specified track. The IBM container is sensed and filled with the correct amount of blend. The operator places the valve housing assembly on top of the full IBM container and secures it in position. All of the impregnation components, such as the marked cap and grip, the filled IBM, the glued parts, and the coded duplex cartons, are positioned near the conveyor. An operator enters a named grip onto the conveyor's moving pugs, followed by another operator inserting a filled IBM into the identified grip.

The component is then impregnated with the impregnation roller on the conveyor. The impregnation weight is randomly evaluated for appropriateness. An operator then attaches a designated cap on the impregnated component. The component is next routed via the coding machine, where the manufacture date, batch number, and variant are printed.

An operator picks up the coded component, inspects it for defects, and places it in a coded duplex carton, which is subsequently delivered to the sealing machine. Each duplex carton should hold the necessary quantity of shoe shiners. The L-Sleeve operator fits the duplex carton into the shrink sleeve and sleeves it. The sleeved duplex carton is then moved to the conveyor in the sleeve tunnel. At the opposite end of the shrink tunnel conveyor, an operator receives the sleeved duplex carton, inspects it for faults, and places it inside the carton.

6. Results and Discussion

Simulation analysis can foresee the unexpected impacts of changes to the work environment and allow for experimenting with alternative scenarios to maximise the organisation's outcomes. In this study, we used discrete event simulation analysis to test several situations. However, in order to grasp the overall process of the current environment, the production manager must first understand the As-is scenario. As a result, we run simulations for the As-is scenario with day one and 8 hours per shift, taking into account performance parameters such as makespan, throughput time, and work in progress. Figure 1 illustrates the makespan data for the entire process on day 1, whereas Figures 2, 3, and 4 depict bar graphs of throughput, work in progress, and throughput, respectively.

The Gantt Chart above represents a one-day simulation of both manufacturing lines, including foam cutting 1 and foam cutting 2, which are operational from 8 a.m. to 11:15 a.m. to cut the foam to the necessary dimensions shown by the blue bar. This operation leaves the production

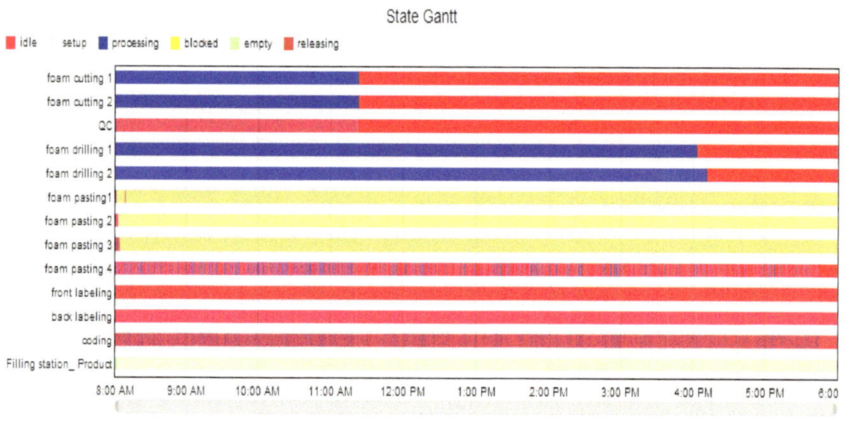

Figure 2. Gantt Chart for Makespan of 1 day simulation of As-is model.

lines idle. Between 8 and 11:15 a.m., shown in Figure 1, personnel are constantly inspecting for air holes in each shoe shiner foam. Between 8 and 11:15 a.m., the quality inspection personnel in Figure 2 are constantly inspecting for air holes in each shoe shiner foam, and reject the pieces if they do not follow the standard, hence it takes a long time among all the processes and acts as a bottle neck for subsequent processes.

In the foam drilling procedure, two professional individuals are recruited to manually place the foam pieces on the pugs of the automated drilling equipment. Again, two machines (hot wire coils) are operated in a row, one from 8 a.m. to 4 p.m. and the other from 8 a.m. to 4:10 p.m.

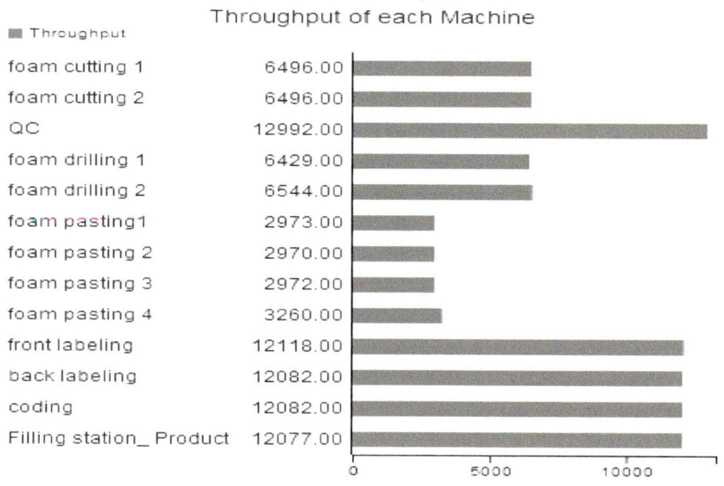

Figure 3. Throughput of each machine per hr of each station for 1 day simulation.

Following this task, the drilling machines are idle. The pasting technique begins shortly after drilling. This image shows two machines, each with two nozzles. The pasting process uses four nozzles from 8 a.m. to 6 p.m. as shown in the yellow bar in Figure 1. Two automated machines work alongside manual intervention to label the front and back beginning in the morning and continuing until 6 p.m. in the evening, following which batch coding is performed.

We looked into the assembly line manufacturing difficulty in the shoe polish industry, where activities are semi-automatic. They struggle to meet demand during high seasons with two cutting machines and two hoppers in scenario 1, with a daily throughput of 12077 shown in Table 1. To locate the bottleneck, we investigated the methods, machines, equipment, manpower, and time involved in each process, and the complete production line was simulated for several types of operations using Flexim simulation software.

Table 1. Results of 5 scenarios and their output.

Scenario\ Parameters	Throughput (per hr)	Throughput (per day)	Throughput (for 26 days)	Technology	Equipment
Scenario 1 (As-is)	1207	12077	314002	Semi-automatic	2 cutting m/c's 2 hopers
Scenario 2	1256	12560	326560	Semi-automatic	1 cutting m/c's 2 hoopers
Scenario 3	1256	12560	326560	Semi-automatic	2 cutting m/c's 3 hoopers
Scenario 4	1350	13500	351000	Semi-automatic	1 cutting m/c's 3 hoopers

As a result, we created and constructed scenarios by modifying the number of equipment, machines, and workers participating in each production line, as well as making changes to station operations. Redesigning the improvement scenarios indicates that when two cutting machines and two hoppers are used in scenario 1, throughput per day is 12077; when one cutting machine and two hoppers are utilised in scenario 2, throughput per day is somewhat higher at 12560.

However, the throughput remained constant when two cutting machines and three hoppers were used in the shoe polish production line. When one cutting machine and three hoppers were hired, the throughput increased significantly. The simulation program output showed a throughput of 13500. As a result, one cutting machine and three hoppers were recommended as part of the overall system enhancement. Furthermore, doing this will reduce and remove product bottlenecks in meeting demand during peak seasons.

7. Conclusion

To address the restrictions of meeting worldwide demand, XYZ, a shoe polish manufacturer operating in a semi-automated environment, intends to increase output by enhancing current performance. However, planning for the future is difficult without first identifying the bottlenecks in existing processes. As a result, in this study, we thoroughly evaluated the existing environment, developed key performance indicators based on the production manager's expectations, and then investigated the As-is model using a simulation study. According to the findings, the quality check takes a lengthy time, which causes subsequent processes to be delayed. Other observations include the requirement for two cutting machines rather than one, as well as the maximum amount of idle time. Higher improvements are possible with semi-automated systems, where extra hoopers can help to speed up the operation, hence enhancing system performance. We measured system performance using makepan, cycle time, throughput time, and work in progress. The most promising scenarios were selected based on the production manager's needs for future use. This study aids the industry by identifying bottlenecks, enhancing current processes, and advising the production manager on future setup.

References

Barosz, P., Gołda, G., and Kampa, A. 2020. Efficiency analysis of manufacturing lines with industrial robots and human operators. *Applied Sciences*, *10*(8), 2862.

Brecher, C., Epple, A., Fey, M., Königs, M., Neus, S., and Wellmann, F. 2017. Lernende Produktionssysteme. *In*: *Proceedings of the Internet of Production für agile Unternehmen: AWK Aachener Werkzeugmaschinen-Kolloquium*, Vol. 18 .

Chen, J., Hu, P., Zhou, H., Yang, J., Xie, J., Jiang, Y., Gao, Z., and Zhang, C. 2019. Toward intelligent machine tools. *Engineering*, *5*(4), 679–690.

ElMaraghy, H., Monostori, L., Schuh, G., and ElMaraghy, W. 2021. Evolution and future of manufacturing systems. *CIRP Annals*, *70*(2), 635–658.

Fan, D., Yang, L., Wang, Z., Triantafyllou, M.S., and Karniadakis, G.E. 2020. Reinforcement learning for bluff body active flow control in experiments and simulations. *Proceedings of the National Academy of Sciences*, *117*(42), 26091–26098.

Guzman, E., Andres, B., and Poler, R. 2022. Models and algorithms for production planning, scheduling, and sequencing problems: A holistic framework and a systematic review. *Journal of Industrial Information Integration*, *27*, 100287.

Jandaghian, Z., and Berardi, U. 2020. Comparing urban canopy models for microclimate simulations in Weather Research and Forecasting Models. *Sustainable Cities and Society*, *55*, 102025.

Johnson, D., Chen, G., and Lu, Y. 2022. Multi-agent reinforcement learning for real-time dynamic production scheduling in a robot assembly cell. *IEEE Robotics and Automation Letters*, *7*(3), 7684–7691.

Li, Y., Gu, W., Yuan, M., and Tang, Y. 2022. Real-time data-driven dynamic scheduling for flexible job shops with insufficient transportation resources using hybrid deep Q networks. *Robotics and Computer-Integrated Manufacturing*, *74*, 102283.

Liang, M., Cao, P., and Tang, J. 2021. Rolling bearing fault diagnosis based on feature fusion with parallel convolutional neural network. *The International Journal of Advanced Manufacturing Technology, 112*(3), 819–831.

Liu, C., Zheng, P., and Xu, X. 2023. Digitalisation and servitization of machine tools in the era of Industry 4.0: A review. *International Journal of Production Research, 61*(12), 4069–4101.

Luo, S., Zhang, L., and Fan, Y. 2021. Real-time scheduling for dynamic partial-no-wait multi-objective flexible job shops by deep reinforcement learning. *IEEE Transactions on Automation Science and Engineering, 19*(4), 3020–3038.

Parente, M., Figueira, G., Amorim, P., and Marques, A. 2020. Production scheduling in the context of Industry 4.0: Review and trends. *International Journal of Production Research, 58*(17), 5401–5431.

Peruzzini, M., Grandi, F., and Pellicciari, M. 2020. Exploring the potential of Operator 4.0 interface and monitoring. *Computers and Industrial Engineering, 139,* 105600.

Saastamoinen, K., and Maunula, K. 2021. Usefulness of flight simulator as a part of military pilots training–case study: Grob G 115E. *Procedia Computer Science, 192,* 1670–1676.

Shahrabi, J., Adibi, M.A., and Mahootchi, M. 2017. A reinforcement learning approach to parameter estimation in dynamic job shop scheduling. *Computers & Industrial Engineering, 110,* 75–-82.

Wang, L., Zhou, S., Liu, Q., and Huang, Y. 2023, November. Joint Simulation of High Voltage Line Fault Based on ATPdraw and MATLAB. *In: 2023 3rd International Conference on New Energy and Power Engineering (ICNEPE),* pp. 800–805 . IEEE.

Xu, X. 2017. Machine Tool 4.0 for the new era of manufacturing. *The International Journal of Advanced Manufacturing Technology, 92,* 1893–1900.

Yazdani, M., Kabirifar, K., Fathollahi-Fard, A.M., and Mojtahedi, M. 2021. Production scheduling of off-site prefabricated construction components considering sequence dependent due dates. *Environmental Science and Pollution Research,* 1–17.

Zawadzki, P., and Żywicki, K., 2016. Smart product design and production control for effective mass customization in the Industry 4.0 concept. *Management and Production Engineering Review.*

Zhang, N., Liu, Z., Qiu, C., Hu, W., and Tan, J. 2020. Optimis ing assembly sequence planning using precedence graph-based assembly subsets prediction method. *Assembly Automation, 40*(2), 361–375.

Zhang, S., Tang, F., Li, X., Liu, J., and Zhang, B. 2021. A hybrid multi-objective approach for real-time flexible production scheduling and rescheduling under dynamic environment in Industry 4.0 context. *Computers & Operations Research, 132,* 105267.

Zheng, P., Zhang, P., Wang, J., Zhang, J., Yang, C., and Jin, Y. 2022. A data-driven robust optimization method for the assembly job-shop scheduling problem under uncertainty. *International Journal of Computer Integrated Manufacturing, 35*(10–11), 1043–1058.

8

Optimising Box Manufacturing Operations in the Textile Industry: A Simulation-Based Approach for Process Improvement

Box plants play a pivotal role in industries ranging from consumer goods to manufacturing, where efficient packaging is critical to meeting global shipping standards and customer demands. As regulations emphasis e the use of lightweight, durable, and recyclable corrugated boxes for transporting both perishable and non-perishable goods, manufacturers must enhance operational efficiency to remain competitive. This chapter focuses on optimising resource utilisation and identifying bottlenecks in the manufacturing process of a box plant that produces packaging for drawn textured yarn (DTY). Using FlexSim simulation software, we developed a process improvement framework that analyses production throughput, cycle times, machine utilisation, and labour interaction with critical machinery. Key bottlenecks were identified in stages, such as the operation of the Rotary Plate Cutting Machine and Kraft Waste Shredder, where excessive idle times and material handling delays were observed. Automation was introduced in critical production stages, including folding, bonding, and material transport, to improve machine efficiency and reduce work-in-progress (WIP) accumulation. Our analysis demonstrated a reduction in throughput loss, increased system output, and minimised machine downtime, leading to a 26% improvement in overall production efficiency.

1. Introduction

The packaging industry has experienced rapid growth in recent years, driven by several factors including increased e-commerce activity, urbanisation, sustainability initiatives, government policies, and technological advancements (Shi et al., 2024). The global shift towards more environmentally friendly packaging solutions has also fuelled this expansion, as businesses seek recyclable and sustainable materials to align with environmental regulations and consumer preferences. This surge has led to a significant rise in the establishment of box manufacturing facilities, which are critical to various industries, especially those engaged in large-scale production and distribution (Transchel et al., 2024).

A box manufacturing facility, often referred to as a corrugated box manufacturing plant, produces boxes from paper and pulp for packaging, transporting, and handling raw materials or finished products. The demand for such packaging solutions is growing as companies increasingly prioritise secure, lightweight, and cost-efficient ways to transport goods, particularly in the e-commerce and textile sector, which has seen exponential growth post-pandemic.

The manufacturing process begins with the transport of raw materials to the pulp manufacturing facility. Here, pulp made from softwood, hardwood, and recycled fibre s is used to produce paper. The integration of recycled materials in the production process not only helps in reducing costs but also contributes to sustainable production practices. Once the paper is produced, it is transferred to a converting facility where corrugated boxes are made using two types of paper—liner and medium—combined into three layers of corrugated sheets. Corrugated board is shaped into the desired dimensions and configurations. Corrugated cardboard, first used for packaging in 1897 (Talib et al., 2009), remains widely utilised today (Van Hung et al., 2010) due to its lightweight, low-cost, and recyclable properties, making it a staple in the global packaging industry.

Box production involves several technical processes, including unloading, board production, printing, slotting, stitching, loading, and waste material handling. Each stage requires specialised equipment (Freddi and Salmon, 2019). Achieving full automation of the cardboard box production process necessitates mechanisation of a broad range of technological steps. In recent years, advancements in automation have significantly improved production line efficiency, reducing labour costs and error rates. The most challenging stage is folding and bonding, as boxes vary in size and shape (Zhang, 2022). Depending on the design, boxes can be assembled using staples, glue, or without additional fastening materials. Some are built directly from a flat pattern, while others require glueing only one seam to form a closed shape. These flat-packed boxes are easy to store and transport, saving space in storage and shipping, which is an increasingly important factor as warehousing and shipping costs rise.

Various designs and shapes of boxes require specialised equipment such as folding and glueing units. These units are designed to automate the glueing process for cardboard and corrugated boxes, optimising efficiency and ensuring quality (Twede et al., 2014). Properly designing these processes is crucial for maximising operational benefits. As production demands increase, companies must also adhere to stringent quality standards, as deviations in box dimensions or structural integrity can lead to increased wastage and operational inefficiencies. Therefore, the identification of bottlenecks and optimization of the production process are critical for maintaining a competitive edge and ensuring the smooth flow of operations.

In this chapter, we examine the box manufacturing operations of a textile industry during its peak season, when demand for DTY surges. The increased yarn production during this period drives the need for more packaging boxes, leading to heightened demand in the box manufacturing process. Maintaining quality standards while meeting this increased demand is essential to prevent variations, imperfections, wastage, and disruptions in the process flow.

To address these challenges, it is necessary to adopt advanced process optimization techniques. Simulation-based approaches have become increasingly popular for optimising complex manufacturing systems. In particular, Flexim simulation software allows for the development of various production scenarios, helping to identify bottlenecks at each stage of production (Lewicki et al., 2024). By simulating different scenarios, it becomes easier to test process improvements without disrupting actual operations. This enables manufacturers to achieve optimal machine and process efficiency while minimising throughput time and production costs. The objective of our study is to optimise the box production process and reduce throughput time. A simulation environment has been created of the DTY–Box plant and through several scenario based analysis bottleneck stations are identified at each stage of production, thus improving machine and process efficiency.

The rest of the chapter is structured as follows: Section 2 presents the literature review, Section 3 outlines the case study, Section 4 discusses the methodology and detailed process analysis, and Section 5 provides the results and discussion.

2. Literature Review

Optimisation in box plant operations has been the focus of numerous studies, with an emphasis on improving efficiency, reducing waste, and balancing production lines. Various methodologies, from simulation to advanced optimisation techniques, have been applied to achieve these goals. Nouria et al. (2015) designed a monitoring system using structural analysis and bond graph models to enhance fault detection and localizability in a corrugated board factory. Similarly, Wu et al. (2018) employed Flexsim simulation software to optimise production lines and improve machine efficiency, offering practical solutions for balancing processes in box plants. Yang and Liu (2022) further demonstrated the power of Flexsim and the ECRS rule (Eliminate, Combine, Rearrange, Simplify) in identifying bottlenecks and significantly improving production line balance.

In terms of environmental impact, Mourad et al. (2014) used life cycle assessment to assess plant modernisation, revealing notable reductions in global warming potential for cardboard and paper production. This highlights the dual benefits of operational efficiency and sustainability for box plant modernisation efforts. Advanced optimisation techniques have also been explored. Baykasoglu et al. (2024) proposed a matheuristic approach that integrates mixed-integer linear programming (MILP) with Simulated Annealing to determine optimal dimensions, production requirements, and purchasing amounts for corrugated boards. Their method minimis ed waste costs and stock diversity, key considerations for efficient box plant management. Lidberg et al. (2020) combined discrete event simulation with multi-objective optimisation to reduce lead times and inventory levels while maintaining high delivery precision, which is critical for managing demand fluctuations in box plants.

Process improvement strategies have also been explored. Keyser et al. (2022) reduced setup times on rotary die-cutters by 42% using the Single-Minute Exchange of Dies (SMED) methodology, combined with control charts and the Five Whys technique, significantly improving throughput in box production. Automation in packaging processes, explored by Zhai et al. (2011), showcased the potential of fully automated systems to improve production efficiency and reduce labour intensity, a system adaptable to box plant operations. Kliment et al. (2022) demonstrated the utility of digital modelling using FlexSim software, allowing companies to simulate and optimise production processes virtually before implementing changes in real life. This approach reduces trial-and-error and helps improve overall process efficiency. To provide a clearer overview, a summary of key studies in this field is presented in Table 1. These studies collectively highlight the importance of integrating advanced optimisation, simulation tools, and automation systems to enhance efficiency, reduce waste, and improve operational sustainability in box plant operations.

Table 1. Summary of Key Studies on Box Plant Operations.

S. No.	Article	Objective	Methodology	Outputs
1	Nouria et al. (2015)	Design a monitoring system for industrial glue production.	Structural analysis, bipartite graphs, bond graphs.	Simulation results on fault detectability and localizability.
2	Mourad et al. (2014)	Assess benefits of plant modernis ation for CB/FBB and KP production.	Cradle-to-gate life cycle assessment.	51% reduction in GWP for cardboard, 9% for paper.
3	Baykasoglu et al. (2024)	Optimis e dimensions and purchasing of corrugated boards.	MILP model integrated with Simulated Annealing.	Computational results showing minimal stock diversity and waste.
4	Wu et al. (2018)	Optimis e production line without expansion.	Flexsim simulation software.	Statistical data showing improved process capability.
5	Liu et al. (2022)	Analys e and optimis e production line balance.	Flexsim simulation, ECRS rule.	Significant improvement in production line balance.
6	Lidberg et al. (2020)	Reduce lead time and storage while maintaining delivery precision.	Discrete event simulation, multi-objective optimization.	Simulation data, cluster analysis, decision rules.
8	Keyser et al. (2022)	Reduce setup time on rotary die-cutter by 25%.	SMED methodology, Five Whys, control charts.	Setup time reduced by 42%.
9	Zhai et al. (2011)	Improve production efficiency with automated carton packing.	STM32 microcontroller for automation and monitoring.	Full automation and increased efficiency in industrial production.
10	Kliment et al. (2022)	Create a digital model of the production line.	FlexSim software module.	A digital model for optimis ing production processes.

3. Case Study

XYZ Company, established in 1989, located in western India, has grown to become a leading player in the global packaging industry and operates a carton box manufacturing plant that is crucial to its supply chain network. With over three decades of expertise, the company boasts a robust turnover of INR 500 crore, reflecting its strong market presence and financial stability. Employing a workforce of 2,443 skilled professionals, XYZ Company prides itself on delivering high-quality, innovative packaging solutions to a diverse client base. Its state-of-the-art carton box manufacturing plant is a testament to its commitment to cutting-edge technology and operational excellence, positioning it as a trusted partner in the packaging sector. The box plant serves as a dedicated facility for supplying boxes to package finished goods for various other plants within the XYZ company. The entire box plant production process has been divided into six stages for detailed analysis. These stages include the Unloading Process, Board Production Process, Printing and Slotting Process, Stitching Process, Loading Process, and Waste Material Handling Process. Each process stage plays a vital role in ensuring the production of high-quality boxes that meet the dimensional and structural standards. The quality is highly valued since it is used to pack the final products of the downstream plants within. As a result, there is a direct relationship between box quality issues and subsequent operations, specifically the dimensions, structural integrity, or alignment. Failure to meet the required quality can also affect the production process flow and lead to inefficiencies. The boxes are required to be close to perfect with strict quality controls, including specific dimensions of width, height, thickness, strength, and the ability to adhesion between them when folding them. Moreover, print quality and stitching uniformity are significant in the production process because these qualities affect the way the boxes are handled and packed with the finished products.

This case study focuses on the application of FlexSim simulation to optimise operations at a box manufacturing plant that produces carton boxes. The plant layout comprises key machines such as Single Facer, Double Facer, Numerical Control Cutter, Stacker Unit, Printing Machines, Auto-Stitching Machines, and Bailing Machines, all working sequentially to produce finished boxes. An in-depth analysis was conducted on the plant's throughput, cycle time, machine utilisation, and layout efficiency. The Gantt chart analysis provides a comprehensive visualisation of machine states, highlighting periods of idleness and blockage. The simulation proved to be a valuable tool for identifying bottlenecks and optimising production flow, enabling data-driven decisions for future capacity planning and process improvement. This project demonstrates the importance of real-time simulation in manufacturing, particularly in identifying inefficiencies and implementing strategic solutions to improve productivity and operational efficiency. This case study highlights the importance of using advanced simulation techniques to drive continuous improvement and achieve long-term operational success.

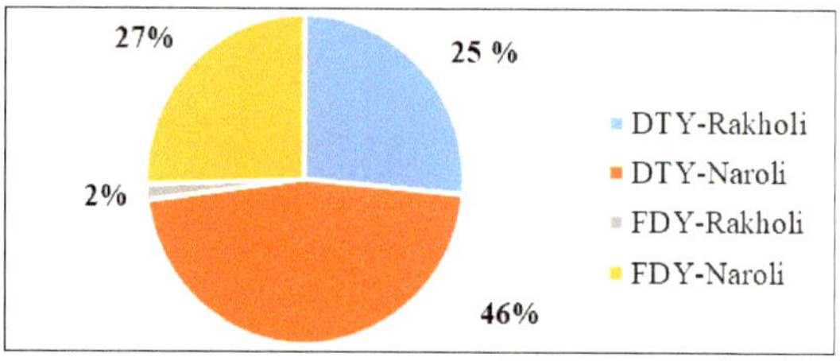

Figure 1. Plant-wise box consumption.

It is observed that box consumption is required in four major units, as shown in Figure 1. The DTY unit at Naroli has the highest consumption of 46% of the total boxes. After this, in the Naroli plant, the FDY (Fully Drawn Yarn) unit required 27%, and the DTY unit in Rakholi desired 25% of the total boxes. The FDY unit of Rakholi has shown that it needs only 2% of boxes. According to the shared data, the daily average production in the day shift (12 hr) is around 70000 boxes and in the Night shift, 40000 boxes, whereas the required number is around 96,000. Typically, the plant operates 12-hour shifts daily, six days a week, from Monday to Saturday. Sundays are generally reserved for maintenance activities. However, in exceptional cases, production may extend to night shifts and Sundays to accommodate increased demand or specific production requirements. Based on the observation, we considered the yield rate of 94–95% for the purpose of this analysis. Also, this value will slightly change with respect to the three- and five-layer boards and the box slotting requirements.

Data

The primary raw materials required in the box plant include Kraft paper, Ink, Corrugation Powder, Stitching Wire, and Modified starch (gum powder). These are basic inputs to the manufacturing process, where kraft paper is used as the primary material for the production of the box, ink for printing, corrugation powder for strength to the board, stitching wire to assemble the box, and gum powder for adhesive.

In addition, the box plant needs other materials to process the production support. The caustic soda is used in cleaning and maintenance, strapping patti and high-temperature tissue tape required in packaging finished boxes and wire used to handle the waste ba ling machines. All

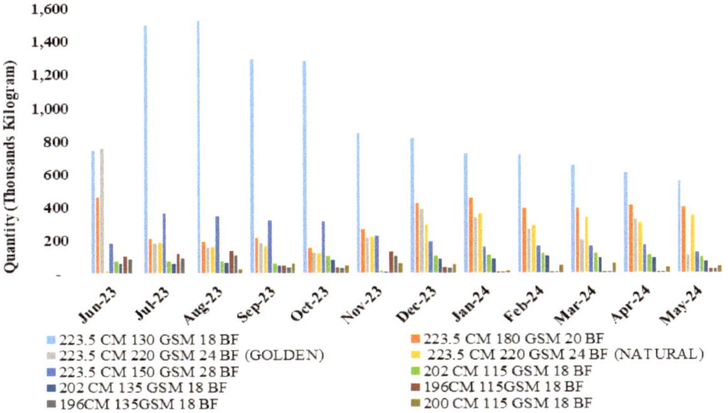

Figure 2. Monthly consumption of top 10 Kraft paper.

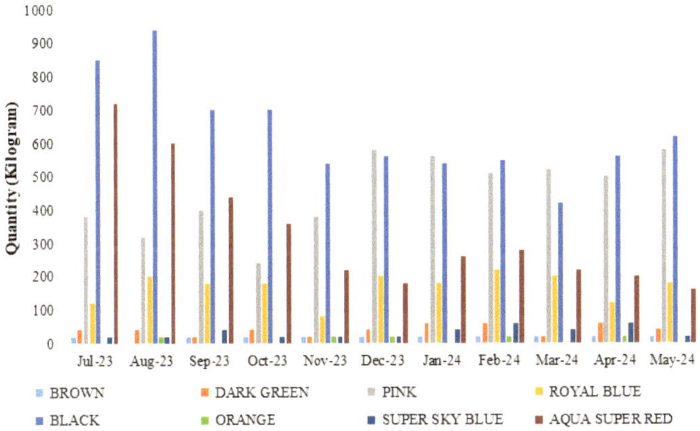

Figure 3. Monthly Consumption of Ink Water Base.

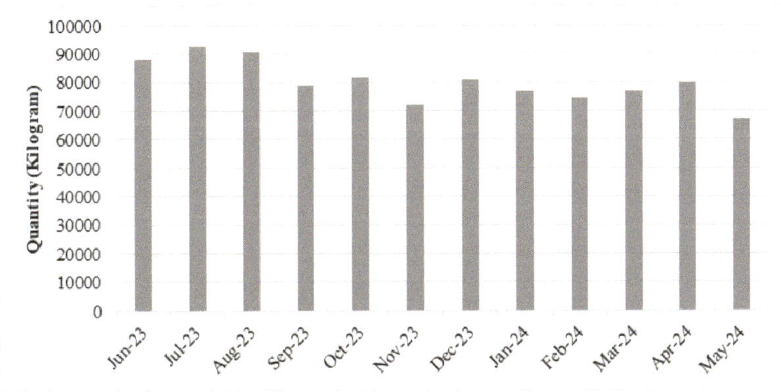

Figure 4. Monthly Consumption of Corrugation Powder.

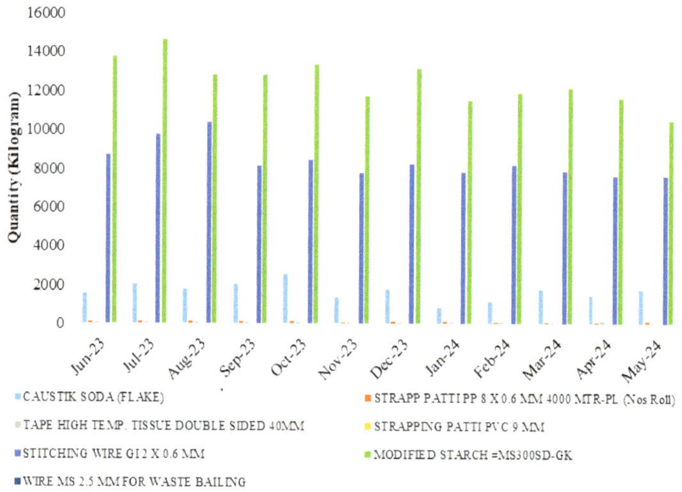

Figure 5. Other raw material monthly consumption.

these raw materials help to facilitate the production of the boxes while meeting the required quality standards.

4. Methodology

The process flow diagram of a box plant consists of six stages, from raw material handling to final product dispatch. The process begins with the unloading of raw materials using forklifts, followed by the storage of these materials in godowns. The board production processes the materials through Single Facers and feeds them into a Double Facer Unit. Adhesives are prepared in the Glue Kitchen and integrated into the board production line. The formed boards are then cut using a Numerical Control Cutter and organised via a Stacker Unit.

The production flow continues into the Printing Process, where a conveyor transports the boards to Printing Machines. Following printing, the boards move into the stitching process, which auto-stitching machines handle. The Waste Material Handling Process is depicted parallelly, showing the shredding of kraft waste, baling, and involvement of a Rotary Plate Cutting Machine.

Finally, the finished goods are transferred using a Hydraulic Hand Pallet Truck to godowns, preparing for loading and dispatch. Figure 6 illustrates the box plant process flow diagram, detailing each stage from raw material handling to final dispatch.

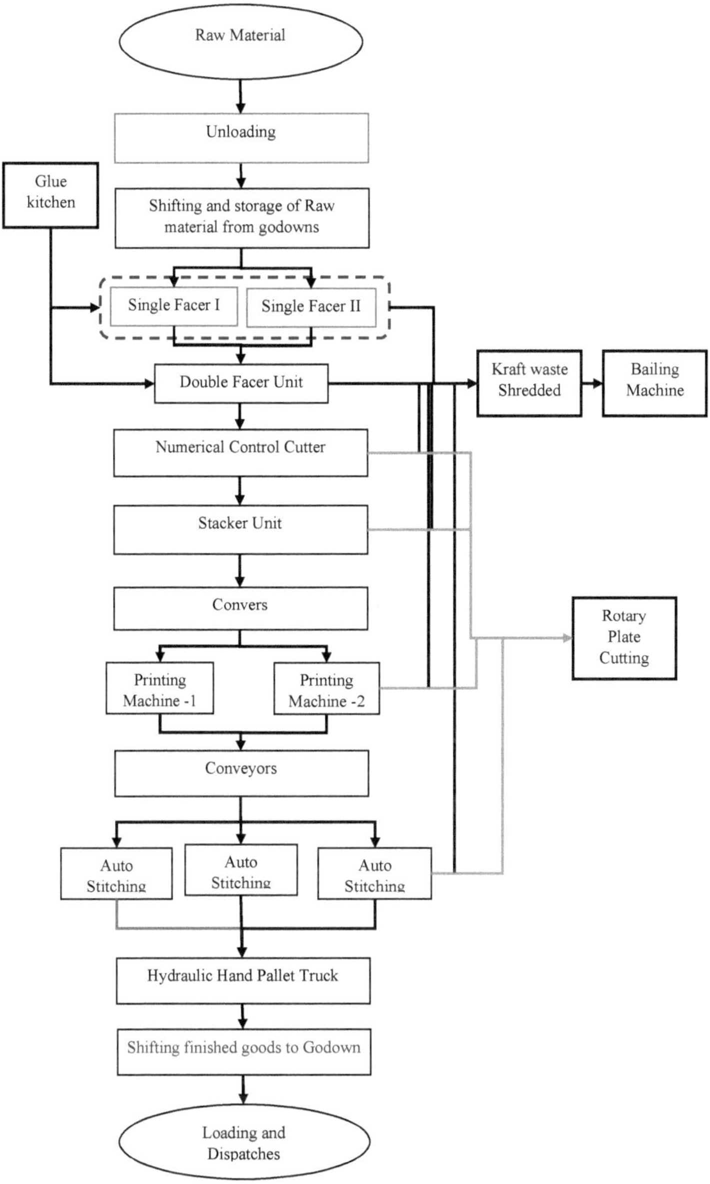

Figure 6. The Box Plant Process Flow.

Process Analysis

This section presents a detailed analysis of the box plant process, focusing on the performed activity and interaction between the worker (Operator and Helper) and the Machine. This analysis was conducted using shared data and based on observations during the plant visit. The analysis aims to identify potential bottlenecks and areas for process improvement to enhance overall productivity and the possible scope of automation in the box manufacturing process. All six stages of the box production process are shown in Figure 7, and the detailed analysis of each activity and current manpower involved in the process are explained in the subsequent section.

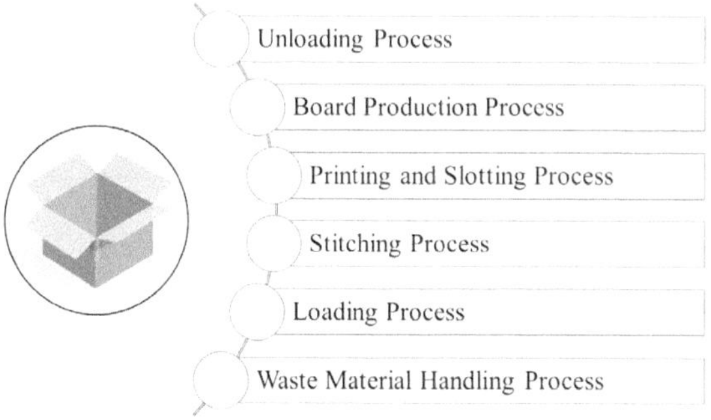

Figure 7. Box Plant Process.

Production Output

As per the shared data from August 2023 to July 2024, we analysed the production volumes of various boxes as finished products. These finished products are categorised by Ply (3-Ply or 5-Ply) and box dimensions in millimetrrs (length × width × height). There are 26 kinds of boxes, whereas only 10 types of boxes (as shown in Figure 8) have more than 20,000 units of average monthly production.

Box type "3 PLY 660 X 440 X 283 MM" is one of the items produced the most, with an average of 5,13,603 units of monthly production. Followed by the "5 PLY 666 X 444 X 280 MM" with 4,98,703 units, "5 PLY 666 X 444 X 283 MM" with 2,44,181 units and "5 PLY 666 X 444 X 244 MM" with 1,86,435 units of monthly average production. The analysis further reveals that the box type "5 PLY 777 X 383 X 244 MM (Expo)" requirement is consistent throughout the year, with a monthly average of 93,195 units. The total yearly production of these boxes is 2,24,25,790 units, whereas the average daily production is 71,878 units, considering 26 working days in months.

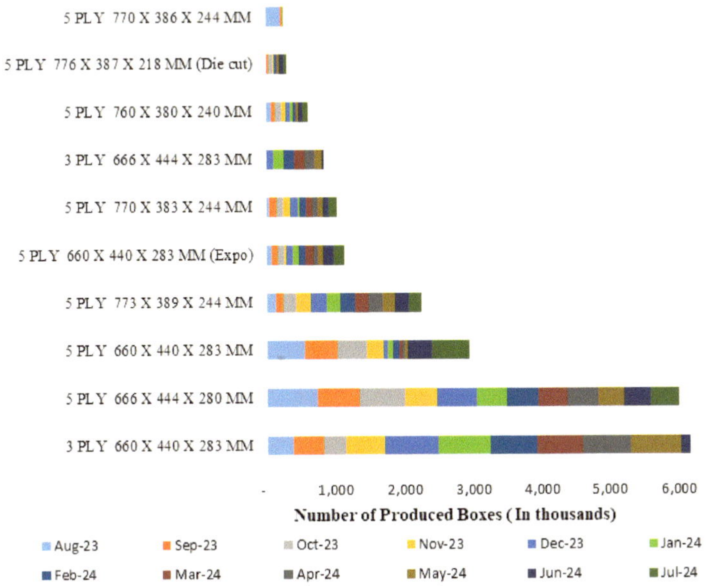

Figure 8. Monthly Production of the Top 15 Finished Products.

Flexsim Demo

FlexSim is an advanced 3D simulation software used for modelling, analysing, visualising, and optimising processes in various industries, including manufacturing, logistics, and healthcare. Its intuitive, drag-and-drop interface allows users to create detailed simulations without extensive coding knowledge. In the live project at XYZ Company, FlexSim was employed to simulate the plant's layout, including key machinery like the Single Facer, Auto-Stitching Machines, and Numerical Control Cutter. By analysing machine throughput, cycle time, and idle states.

The layout of a Box plant features various labelled machines such as "Single Facer", "Double Facer", "Numerical Control Cutter", "Stacker Unit", "Printing machines", "Ba ling machine", and "Auto stitching machine" along with associated queues and trolleys. The layout represents the workflow within the Box manufacturing process.

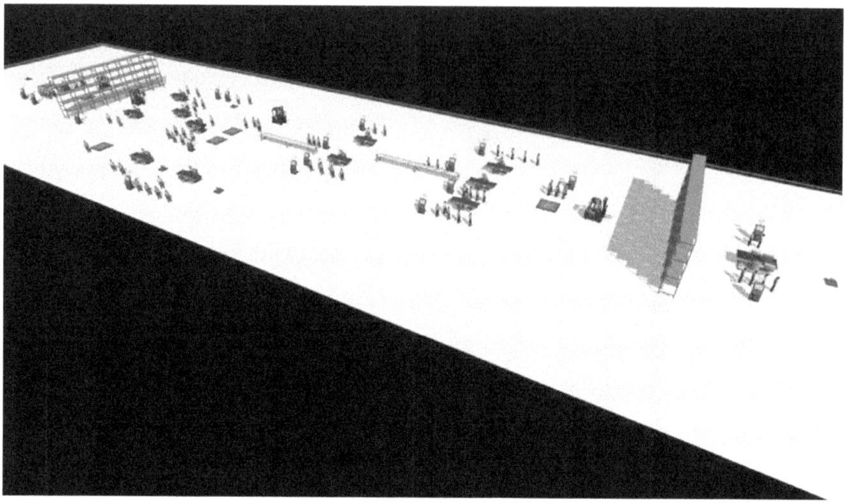

Figure 9. The Box Plant Model.

5. Results and Discussion

This section presents the results of the production system simulation, focusing on key performance indicators to identify bottlenecks and areas for improvement. Analysis includes throughput at both the machine and system levels, a state Gantt chart visualising machine utilisation, and average stay times for each process step. These results aim to provide insights into system efficiency and inform strategies for optimization.

Table 2. Throughput Analysis of the box plant.

Scenarios	Overall Machine Output (1 day)	Overall System Output (1 day)
1	90812 Boxes	67405 Boxes

Table 2 reveals a key insight into the Box plant's production dynamics over a 1-day period. The analysed scenario maintains an identical machine output of 90812 boxes, highlighting consistent production capacity, with a system output of 67405 boxes.

Figure 10 compares throughput at two different levels: machine output and system output. The total throughput of all machines combined is 90812 units, while the final system output is 67405 units. This difference indicates a significant loss of throughput somewhere in the production process between the machines completing their tasks and the final output leaving the shopfloor. This loss could be attributed to several factors, including: work-in-progress (WIP) accumulating between stages, quality

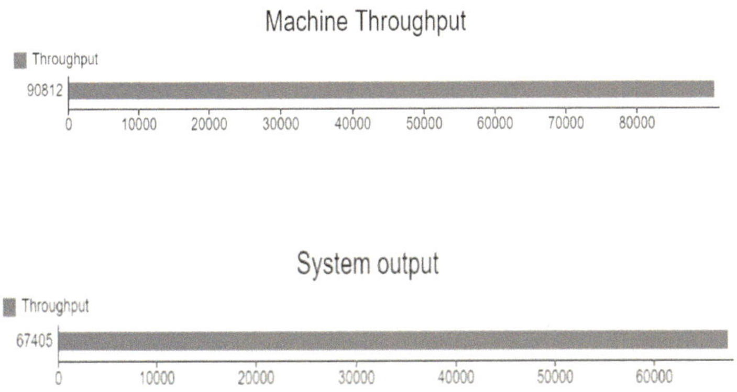

Figure 10. Machine Vs. System Throughput.

control rejections, product rework, inefficient material handling or routing processes, machine downtime for changeovers or maintenance, or even data recording discrepancies. Further analysis is necessary to pinpoint the specific causes of this throughput loss and implement improvements to optimise the shopfloor's overall efficiency and bring its output closer to the total machine processing capacity.

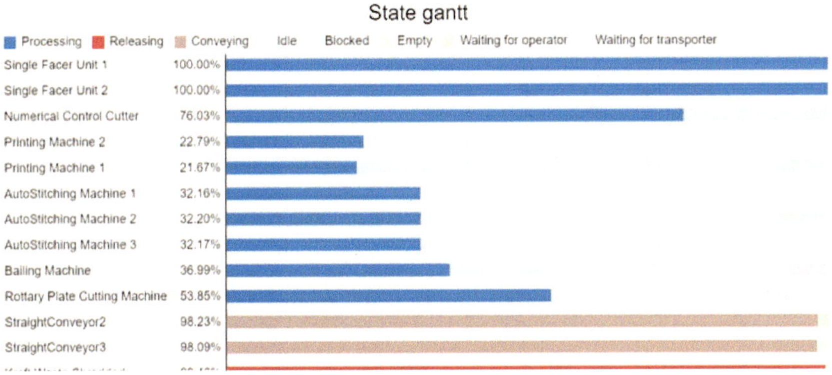

Figure 11. Gantt Chart for Box Plant.

State Gantt chart presented in Figure 11 for a Box Plant, illustrating the operational status of various machines over time. The chart tracks the states of "Single Facer-1&2", "Double Facer", "Numerical Control Cutter", "Printing Machine-1&2", "Auto Stit ching Machine-1,2&3", "Ba ling Machine", and "Rottary Plate Cutting Machine", with colour codes indicating processing, setup, idle, and blocked states. It provides a visual overview of machine utilis ation and potential daily bottlenecks. The Gantt chart reveals significant inefficiencies in the production system, primarily characterised by substantial idle time across machines. While the conveyors and Kraft Waste Shredder operate at near full capacity, their constant activity may mask upstream bottlenecks.

The Cycle time for each box to produce will be 1.66 + 1.66 + 0.71 + 0.5 + 0.54 + 1.01 + 1.11 + 1.15 + 11.80 + 12.44 = 32.58 Seconds/Box. Bar chart illustrated in Figure 12 shows the average stay time of various machines and components in a Box plant. This data provides insights into the time each machine holds the material during processing. The figure suggests that the Kraft Waste Shredded and Rotary Plate Cutting Machine stages are potential bottlenecks in the overall process.

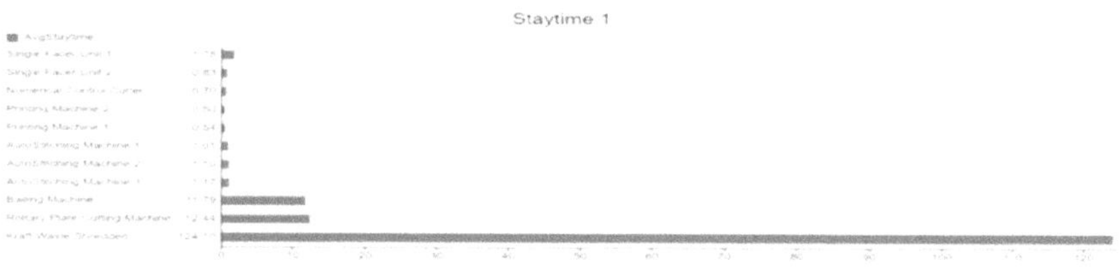

Figure 12. Staytime of the Processors, Transporters & Queues.

6. Conclusion

To meet the growing demand for corrugated boxes in the DTY and FDY packaging units, the semi-automated box production facility aims to enhance its output by improving current performance. However, preparing for future demands requires identifying existing bottlenecks. This study thoroughly analysed the current production environment, defined key performance indicators in collaboration with the production manager, and investigated the As-is model using simulation. The analysis revealed a significant gap between machine capacity and final system output—90812 units produced versus 67405 units delivered—indicating considerable throughput loss. This discrepancy highlights inefficiencies in production, material handling, or final output stages.

Further analysis using Gantt charts pinpointed bottlenecks caused by idle time across various machines. Recommendations were made to improve coordination between production stages and close the gap between machine capacity and system output. Additionally, attention is needed to enhance upstream production by optimising conveyor and waste shredder usage, synchronising material flow, improving routing, and maximising the utilisation of the Rotary Plate Cutting machine. Automation was suggested for certain processes, especially folding and bonding, where variations in size and shape pose challenges. Further improvements were recommended for the quality control and material handling departments, where inefficiencies were observed. Using FlexSim simulation software, various scenarios were modelled to provide insights into process optimization. This helped reduce cycle times, optimise bottleneck machines, minimis e WIP accumulation, and enhance material flow. Preventive maintenance schedules were also advised to reduce machine downtime. Future work may include sensitivity analysis of machine parameters, the introduction of advanced production techniques, and testing of further automation scenarios to achieve higher system efficiency and maximise output.

References

Baykasoğlu, A., Subulan, K. and Karaman, İ. 2024. A matheuristic-based solution approach for integrated stock selection and sizing problem with a case study in a corrugated box-production company. *Computers & Industrial Engineering, 195*, 110429.

Freddi, A., Salmon, M., Freddi, A. and Salmon, M. 2019. Design of a Packaging Machine: General Description and Conceptualization. *Design Principles and Methodologies: From Conceptualization to First Prototyping with Examples and Case Studies*, 205–237.

Keyser, R. S., Severin, R. S. and Geiger, M. J. 2022. Setup time reduction with SMED in a corrugated box plant. *Journal of Applied Research on Industrial Engineering, 9*(2), 264–271.

Kliment, M., Trojan, J., Pekarcíková, M. and Kronová, J 2022. Creation of simulation models using the flexsim software module. *Advanced Logistic Systems-Theory and Practice, 16*(1), 41–50.

Lewicki, W., Niekurzak, M. and Wróbel, J. 2024. Development of a Simulation Model to Improve the Functioning of Production Processes Using the FlexSim Tool. *Applied Sciences, 14*(16), 6957.

Lidberg, S., Aslam, T., Pehrsson, L. and Ng, A. H. 2020. Optimis ing real-world factory flows using aggregated discrete event simulation modelling: Creating decision-support through simulation-based optimization and knowledge-extraction. *Flexible Services and Manufacturing Journal, 32*(4), 888–912.

Mourad, A. L., da Silva, H. L. G. and Nogueira, J. C. B. 2014. Life cycle assessment of cellulose packaging materials production: folding box board and kraft liner paper. *The International Journal of Life Cycle Assessment, 19*, 968–976.

Nouria, T., Larbi, S., Hafid, H. and Belkacem, O. B. 2015, May. Monitoring system for manufacturing process: Application to a cardboard factory. *In: 2015 3rd International Conference on Control, Engineering & Information Technology (CEIT),* pp. 1–6 . IEEE.

Shi, Y., Lin, Y., Wang, S., Wen, H., Yang, L., Lim, M. K. and Li, Y. 2024. A simulation-optimis ation system for recycling logistics network of recyclable express packaging. *Computers & Industrial Engineering, 189*, 109949.

Talbi, N., Batti, A., Ayad, R. and Guo, Y. Q. 2009. An analytical homogenization model for finite element modelling of corrugated cardboard. *Composite Structures, 88*(2), 280–289.

Transchel, S., Taheri-Bavil-Oliaei, M. and Petersen, M. 2024. Reusable packaging for B2C supply chains. *In: Sustainable Supply Chains: A Research-based Textbook on Operations and Strategy,* pp. 405–431 . Cham: Springer International Publishing.

Twede, D., Selke, S. E., Kamdem, D. P. and Shires, D. 2014. *Cartons, Crates, and Corrugated Board: Handbook of Paper and Wood Packaging Technology*. DEStech Publications, Inc.

Van Hung, D., Nakano, Y., Tanaka, F., Hamanaka, D. and Uchino, T. 2010. Preserving the strength of corrugated cardboard under high humidity conditions using nano-sized mists. *Composites Science and Technology, 70*(14), 2123–2127.

Wu, G., Yao, L. and Yu, S. 2018, June. Simulation and optimization of the production line based on FlexSim. *In*: *2018 Chinese Control And Decision Conference (CCDC),* pp. 3358–3363 . IEEE.

Yang, Q. and Liu, W. 2022. Production line balance optimization design of S company. *Manufacturing and Service Operations Management, 3*(1), 57–71.

Zhai, Y., Wang, L., Liu, Y., Jiang, T. and Jiang, H., 2021, June. The Design of Automatic Carton Packing System of Tobacco Based on STM32. *In*: *Journal of Physics: Conference Series* (*1948* (1), 012099). IOP Publishing.

Zhang, X. 2022, February. The Application of Folding Structure in Paper Packaging Design. In *2021 Conference on Art and Design: Inheritance and Innovation* (*ADII 2021*), pp. 281–285. Atlantis Press.

9

Enhancing Paper Tube Production in the Textile Industry: A Simulation-Driven Strategy for Process Improvement

This chapter investigates the operational efficiency of a paper tube manufacturing plant that supports the textile industry, focusing on challenges encountered during peak production periods. As global demand for sustainable packaging materials rises, particularly in sectors like textiles, paper tube manufacturers must meet increased orders without sacrificing product quality. The chapter examines the key operational bottlenecks in the manufacturing process and explores strategies to mitigate these issues, ensuring smooth downstream operations and minimising wastage. A significant part of the analysis includes using simulation tools like FlexSim to model production scenarios, identify inefficiencies, and optimise machine utilisation. Through process optimization, the plant achieves its production targets while adhering to stringent quality standards required by the textile industry. The insights provided in this chapter demonstrate how leveraging simulation-based tools and continuous process improvements can significantly impact the efficiency and sustainability of paper tube manufacturing in today's competitive market.

1. Introduction

The global paper industry, a cornerstone of modern commerce and communication, has experienced remarkable growth in recent decades, fueled by the burgeoning demand for packaging, hygiene products, and sustainable alternatives to plastic (Dai et al., 2024). This surge, evidenced by the Food and Agriculture Organization of the United Nations (FAO) reporting nearly 400 million metric tons of paper and paperboard produced in 2022, highlights the industry's pivotal role in a rapidly evolving global landscape (FAOUA, 2022). While the rise of e-commerce and digital media has undeniably impacted certain paper markets, it has simultaneously driven an increased need for paper-based packaging materials, particularly within the rapidly expanding e-commerce sector. This dynamic interplay of evolving consumer behaviour and technological advancement underscores the complex nature of the paper industry's trajectory. Currently valued at approximately USD 362 billion (2023), the global pulp and paper market is projected to reach USD 482 billion by 2032, exhibiting a compound annual growth rate (CAGR) of 3.2% (Statista, 2024). This projected growth is largely attributed to the increasing adoption of recycled paper products and the escalating demand for sustainable packaging solutions, especially within the food packaging and e-commerce industries.

The paper industry is divided into several segments, including packaging, printing and writing papers, tissue products, and specialty papers (Crini et al., 2020). Among these, packaging papers and board, including corrugated boxes and paper bags, represent the largest and fastest-growing segment (Reichert et al., 2020). The increasing shift towards sustainable packaging solutions, largely due to consumer awareness of environmental concerns, has led to greater demand for paper-based packaging alternatives (Tawfik et al., 2022). Furthermore, regulatory actions such as plastic bans in several countries have further boosted the industry. For example, Asia-Pacific dominates the global market, with China being the leading producer and consumer of paper products, followed by North America and Europe.

A crucial segment within the paper industry is the paper tube manufacturing sector. These cylindrical containers, also known as paper cores or cardboard tubes, are crafted from recycled paperboard and find diverse applications across various industries, including packaging, textiles, construction, industrial uses, and consumer goods. Paper tubes are formed by winding layers of paper or cardboard around a mandrel, resulting in a versatile, durable, and recyclable product. Their significance lies in their increasing role as a replacement for plastic and metal-based packaging solutions, offering a biodegradable, cost-effective, and environmentally friendly alternative. This sector is vital to the broader paper industry. It supplies essential components for winding materials like yarn, films, fabric, and paper, demonstrating their utility and contribution to sustainable practices.

The demand for paper tubes is experiencing a surge, mirroring the global growth of the packaging and textile industries. Packaging companies, in particular, rely heavily on these tubes to create robust yet lightweight cores for various products. Moreover, the intensifying focus on reducing plastic waste has propelled the adoption of eco-friendly alternatives like paper tubes, further accelerating market growth. Recent market studies project significant expansion for the global paper tube industry in the coming years, driven by its alignment with prevailing sustainability trends. However, this anticipated growth is accompanied by increasing pressure on manufacturers to optimise production processes, striving to meet burgeoning demand while simultaneously controlling costs and minimising waste.

Paper tube manufacturing is a multi-stage process beginning with carefully selecting paperboard, typically sourced from recycled or virgin fibres. This choice impacts the final tube's strength, cost, and environmental footprint. The selected paperboard is then fed into specialised winding machines. These machines precisely wind the paperboard around a mandrel, a cylindrical core, in multiple layers. The number of layers determines the tube's wall thickness, strength, and crush resistance. During the winding process, adhesives are applied to bind the layers together securely, ensuring the structural integrity of the finished tube. The type of adhesive used is carefully chosen based on the intended application of the tube, considering factors like humidity resistance and food safety requirements (Gadhave and Gadhave, 2022). Once the tube is formed on the mandrel, it is cut to the specified length using precision cutting tools. Further processing steps can include adding features like end caps, inner liners, or external coatings to enhance functionality and protect the contents. Printing, often directly onto the tube's surface, can incorporate branding, product information, or decorative elements.

This flexible production process allows manufacturers to tailor paper tubes to a wide range of specifications, making them suitable for diverse industries, including construction (forming concrete pillars), textiles (winding yarn or fabric), and food packaging (holding snacks or powdered products). Leading companies like DS Smith and International

Paper continuously innovate within this sector, developing new materials, processes, and applications to meet evolving market demands and address global sustainability challenges. Despite the inherent adaptability of the process, manufacturers still face ongoing challenges. These include optimising production flow to eliminate bottlenecks, minimising material waste and scrap generated during the cutting and forming processes, and managing adhesive usage effectively to balance performance with environmental impact. These challenges highlight the need for continued research and development in paper tube manufacturing to enhance efficiency, sustainability, and cost-effectiveness.

One of the key challenges in the paper tube manufacturing industry is bottlenecks, which occur when specific steps in the production process slow down the overall workflow. For instance, machines that are responsible for cutting or winding may cause delays if their capacities do not match the upstream processes. These bottlenecks can result in longer production times, higher operational costs, and reduced efficiency. Process improvements, such as better machine synchronisation and automation, can significantly reduce the impact of bottlenecks. Automation helps streamline the production flow by ensuring consistent machine speeds and reducing human error.

In this chapter, we examine the operational efficiency of a paper tube manufacturing plant that supports the textile industry, particularly during peak production periods when the demand for tubes rises sharply. Maintaining quality while meeting increased demand is crucial to minimising wastage and ensuring seamless operations in downstream processes. To optimise the manufacturing process, simulation-based tools like FlexSim are used to model production scenarios, identify bottlenecks, and propose solutions for enhancing throughput and machine utilisation. These optimizations help ensure that the plant can meet its output targets while maintaining the stringent quality standards required by the textile industry.

The rest of the chapter is organised as follows: Section 2 presents the case study, Section 3 details the methodology and process analysis, and Section 4 discusses the results and Section 5 conclusions.

2. Case Study

XYZ Company, established in 1989 and located in western India, has grown to become a leading player in the global packaging industry, with a turnover of INR 500 crore. Over three decades, XYZ has maintained a strong market presence, providing high-quality, innovative packaging solutions to a diverse client base. Employing 2,443 professionals, the company operates a state-of-the-art paper tube manufacturing plant, which plays a critical role in the supply chain by providing essential components to other production units within the company. The plant runs six days a week in 12-hour shifts, focusing on producing durable, high-precision paper tubes essential for downstream manufacturing.

This case study focuses on the operational efficiency at the paper tube plant, highlighting the key process stages and challenges faced in maintaining high productivity. The paper tube manufacturing process is divided into the following critical stages; Raw Material Unloading: Kraft paper, adhesives, and parchment paper are manually unloaded. This labour-intensive process requires significant time due to the size and weight of the materials, with each Kraft paper roll weighing up to 930 kg. **Slitting:** Large rolls of Kraft paper are cut into smaller widths by operators working on slitting machines. **Winding:** Tubes are formed by winding multiple layers of cut paper to achieve the required thickness. **Heat Treatment:** Tubes undergo oven treatment to harden and achieve dimensional stability. **Auto-Finishing:** Tubes are cut, notched, and finished to meet precise specifications. **Packaging and Dispatch:** Finished products are barcoded, packed, and dispatched for use in other production units. The operational challenges of the plant are as discussed. Firstly, manual labour, the unloading of raw materials is a physically demanding task, especially given the weight of the rolls and the need for precision during transport. Secondly, bottlenecks in slitting, slitting is a crucial stage that determines the speed of downstream processes. Delays or errors here can lead to inefficiencies in winding and finishing. Thirdly, quality control, dimensional precision is critical for the tubes, as any deviation affects downstream production. The plant faces strict quality requirements, particularly in tube dimensions and structural integrity. Optimization can be done through simulation to address these operational challenges, a simulation using FlexSim was applied to optimise machine utilisation and throughput. The analysis identified bottlenecks in the slitting and winding stages, with solutions suggested to improve overall efficiency by balancing workforce allocation and enhancing machine scheduling. This strategic approach enabled XYZ to make data-driven decisions for capacity planning, ensuring continuous improvement in the plant's operations.

3. Methodology

The process flow diagram of a tube plant, as shown in the figure 1, consists of several stages, from raw material handling to final product dispatch. The process begins with the unloading of raw materials such as kraft paper and adhesive using appropriate handling equipment. These materials are stored in designated raw material godowns to maintain proper inventory and workflow. Next, the materials are transferred to the plant for slitting, where they are processed through multiple slitting machines. These machines cut the kraft paper and other materials into required sizes. After the slitting process, the paper is sent to winders for further processing. The winders, typically operating in parallel, handle the coiling of the slit paper into jumbo tubes for loading into the ovens.

The production flow continues as the jumbo tubes are heated in multiple ovens to set the material for the next stages. Once heated, the tubes are transferred to the auto-heading machines, which are organised into various units. These machines automatically cap the ends of the tubes to complete the formation process. The waste material handling process operates concurrently, where kraft waste generated during slitting is recycled. The waste is collected, processed, and either reused in the slitting machines or handled through appropriate recycling and disposal mechanisms. Finally, the finished goods, including both full-length and cut-to-size materials, are packed into boxes. These boxes are transferred to finished goods godowns using material handling equipment, preparing them for loading and dispatch. This flowchart outlines the key stages, from raw material intake to the final dispatch of products, ensuring a streamlined and efficient production flow.

The functioning of a tube plant is understood by simulating the Tube Plant model on FlexSim, a powerful simulation tool, to model and analys e the production flow (Figure 2). By simulating the process flow of raw material handling, slitting, winding, heating, and auto-heading in FlexSim, key bottlenecks and inefficiencies in the plant were identified. The simulation provided insights into optimal resource allocation, machine utilization, and process timings, enabling the team to test various production scenarios without disrupting actual operations. FlexSim's visualisation of the workflow helped highlight areas where process improvements could lead to reduced downtime and increased throughput. The insights gained from this simulation were crucial in optimising the tube plant's production flow, reducing waste, and enhancing overall operational efficiency.

4. Results and Discussion

This section presents the findings from the simulation and optimisation of the paper tube manufacturing process, focusing on how operational efficiency can be enhanced to meet the demands of the textile industry, particularly during peak production periods. Key metrics such as throughput, machine utilisation, and cycle time were analysed, revealing

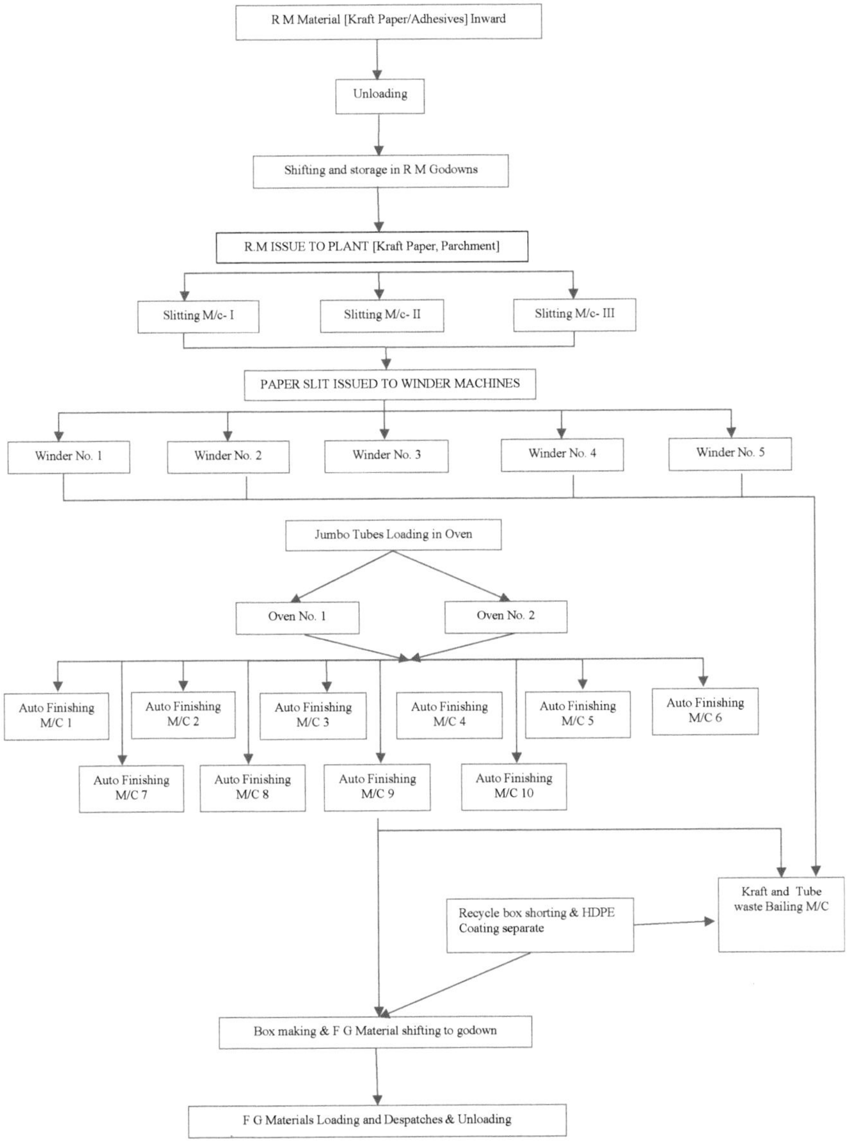

Figure 1. Process flowchart of tube plant.

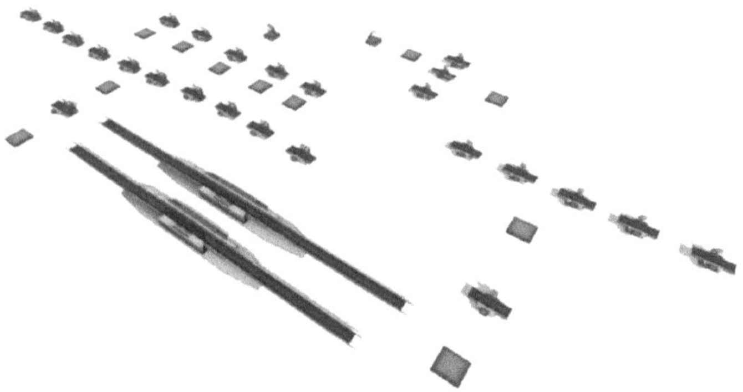

Figure 2. The Paper Tube Plant Model.

critical bottlenecks in production and identifying areas for process improvements. By employing simulation tools like FlexSim, we were able to model different production scenarios and propose solutions for improving performance across the plant. The discussion also contextualis es these results within the broader literature on manufacturing process optimization and highlights the implications for industrial-scale applications, particularly in industries that face cyclical demand patterns.

Table 1 highlights a significant discrepancy between the potential output of individual machines and the actual system output of paper tubes. While the machines are capable of producing 87,663 (12 m) tubes in a day, the system only delivers 12,087 (12 m), indicating a major bottleneck in the production process. This suggests potential issues with machine reliability, quality control, material waste, or production flow inefficiencies. Addressing these underlying problems through process optimisation, improved machine maintenance, and effective quality control measures is crucial to increase overall efficiency and profitability for the paper tube manufacturing operation.

Figure 3 reveals a significant discrepancy between the machine and system output. The figure highlights the total machine throughput, representing the combined output of all machines, at 87,663 units. However, the final system output, reflecting the number of completed tubes leaving the shop floor, is significantly lower at 12,087 units. This difference signifies a substantial loss of throughput somewhere within the

Table 1. Throughput Analysis of the Tube plant.

Scenarios	Overall Machine Output (1 day)	Overall System Output (1 day)
1	87663 Tubes	12087 Tubes

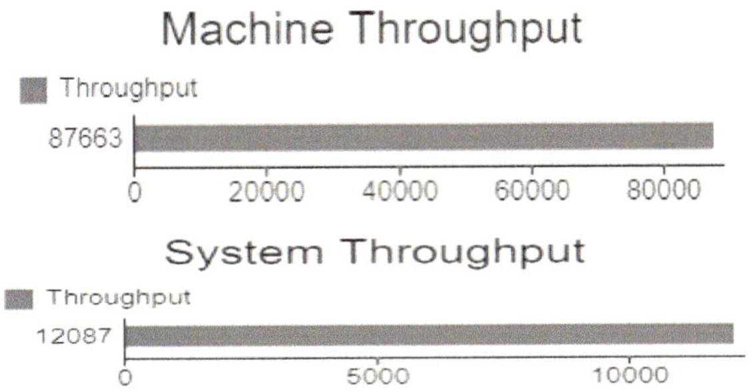

Figure 3. Machine Vs. System Throughput.

production process. Several factors could contribute to this loss, including work-in-progress (WIP) accumulating between stages, quality control rejections, product rework, inefficient material handling or routing processes, machine downtime for changeovers or maintenance, or even data recording discrepancies. Further analysis is crucial to pinpoint the specific causes of this throughput loss. By identifying and addressing

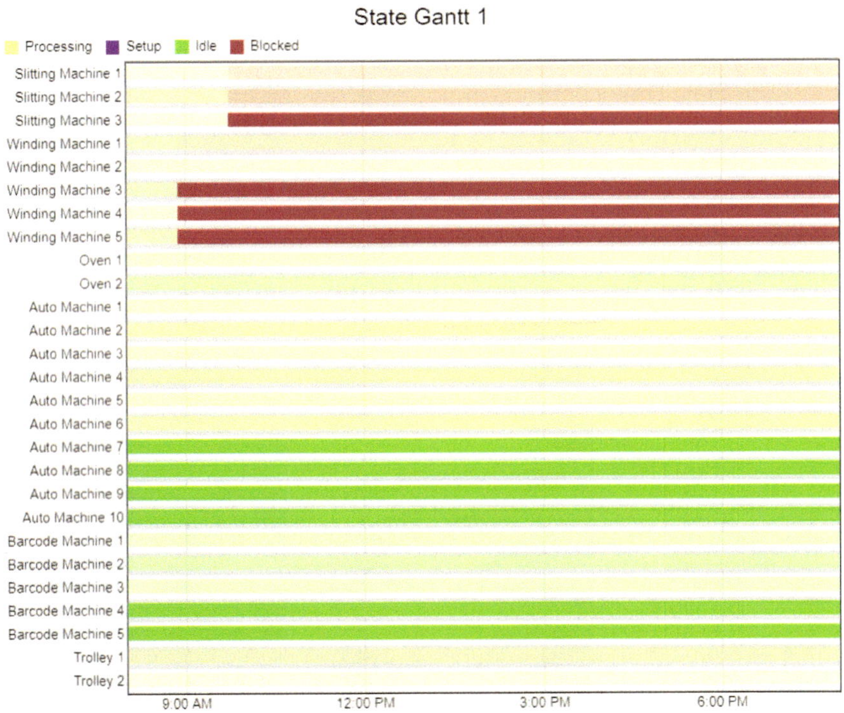

Figure 4. Gantt Chart for the Tube Plant.

these bottlenecks, the shop floor's overall efficiency can be optimised, bringing its output closer to the total machine processing capacity.

The State Gantt chart in Figure 4 presents an overview of the operational states of key machines within the paper tube manufacturing process over a single day. Machines such as slitting, winding, and barcode equipment are monitored in terms of their active processing, setup times, idle periods, and blockages. The results highlight significant operational inefficiencies, particularly among the winding machines and slitting machines. For instance, Slitting Machine 3 and Winding Machines 3, 4, and 5 experienced prolonged periods of blockage (indicated by red), suggesting potential bottlenecks that are disrupting the production flow. These blockages likely result from either resource constraints or upstream delays, contributing to downstream inefficiencies. In contrast, Auto Machines 7 to 10 and the Barcode Machines exhibited high utilisation with minimal blockages, primarily operating in idle (green) or processing (yellow) states. This suggests that while certain parts of the production line are well-optimised, other areas, especially the slitting and winding sections, require process improvements to reduce downtime and increase overall throughput. These findings are critical for identifying and addressing bottlenecks to enhance the plant's operational efficiency during peak demand periods.

The Stay Time chart (Figure 5) presents the average time that each machine or component in the paper tube production process holds a 12-metre tube before it moves to the next stage. This detailed breakdown reveals key contributors to overall cycle time and highlights areas for potential process optimization. The total cycle time per tube is calculated as 53.8 minutes, combining the stay times across the entire production line. From the chart, it is evident that Auto Machine 1 has the longest

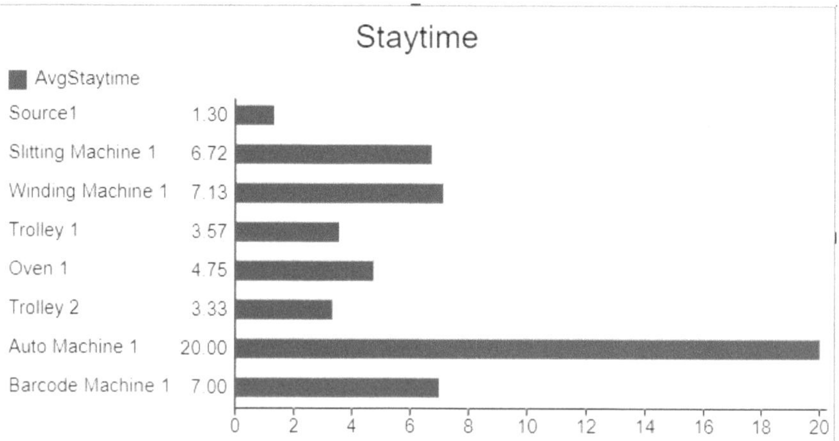

Figure 5. Stay time of the Processors, Trolleys, & Queues.

stay time, holding the material for an average of 20 minutes, representing a significant portion of the total cycle time. Other machines like the Slitting Machine 1 (6.72 minutes), Winding Machine 1 (7.13 minutes), and Barcode Machine 1 (7 minutes) also contribute notable delays. The Oven 1 also adds 4.75 minutes of processing time, reflecting the thermal processes required for tube manufacturing. The two trolleys used for material movement contribute around 6.9 minutes collectively, while the initial Source1 setup has a relatively short stay time of 1.3 minutes. These results pinpoint Auto Machine 1 as a key area for intervention to reduce overall production time, along with opportunities to streamline slitting, winding, and barcode processes. By addressing these bottlenecks through techniques like machine scheduling and process automation, the plant can significantly enhance throughput and reduce the total cycle time, especially during peak production periods.

5. Conclusion

This chapter presents a comprehensive analysis of a semi-automated paper tube manufacturing facility supporting the textile industry, focusing on identifying and mitigating bottlenecks in the production process. The findings revealed a significant gap between machine capacity and system output, which led to considerable throughput loss. By utilising simulation tools such as FlexSim, various production scenarios were modelled to identify inefficiencies and bottlenecks that contribute to idle times and material flow disruptions.

Key areas requiring improvement included the synchronisation between production stages, particularly in upstream processes like slitting, winding, and material handling, as well as the optimisation of machines like the Rotary Plate Cutter. The results also indicated that automating processes, such as folding and bonding, could help address the challenges posed by variations in product size and shape. Moreover, improvements were recommended for quality control and material handling to reduce inefficiencies and enhance overall throughput.

The study's recommendations for implementing preventive maintenance schedules, automating critical processes, and synchronising material flow offer practical strategies for increasing the plant's productivity and closing the gap capacity and output. Future work may explore advanced automation techniques, further between optimization of production flows, and sensitivity analysis of machine parameters to ensure the plant can meet future demand efficiently and sustainably. These process improvements are crucial for maintaining the plant's competitive edge in an industry driven by increasing demand and stringent quality standards.

References

Dai, M., Sun, M., Chen, B., Shi, L., Jin, M., Man, Y., Liang, Z., de Almeida, C.M.V.B., Li, J., Zhang, P., and Chiu, A. S. 2024. Country-specific net-zero strategies of the pulp and paper industry. *Nature, 626*(7998), 327–334.

Global paper industry: Statistics & facts, Statista—2024. https://www.statista.com/topics/1701/paper-industry/

Crini, G., Lichtfouse, E., Chanet, G., and Morin-Crini, N. 2020. Applications of hemp in textiles, paper industry, insulation and building materials, horticulture, animal nutrition, food and beverages, nutraceuticals, cosmetics and hygiene, medicine, agrochemistry, energy production, and environment: A review. *Environmental Chemistry Letters, 18*(5), 1451–1476.

Reichert, C. L., Bugnicourt, E., Coltelli, M. B., Cinelli, P., Lazzeri, A., Canesi, I., Braca, F., Martínez, B. M., Alonso, R., Agostinis, L., and Verstichel, S. 2020. Bio-based packaging: Materials, modifications, industrial applications, and sustainability. *Polymers, 12*(7), 1558.

Tawfik, A., Bakr, M. H., Nasr, M., Haider, J., Lim, H., Qyyum, M. A., and Lam, S. S. 2022. Economic and environmental sustainability for anaerobic biological treatment of wastewater from paper and cardboard manufacturing industry. *Chemosphere, 289*, p. 133166.

Gadhave, R. V. I., and Gadhave, C. R. 2022. Adhesives for the paper packaging industry: An overview. *Open Journal of Polymer Chemistry*, 12(2), 55–79.

Index

About the Authors

Leonilde Varela received her Ph.D. degree in Industrial Engineering and Management from the University of Minho, Portugal in 2007. She is an Associate Professor with Habilitation at the Department of Production and Systems of University of Minho. Her main research interests are in Manufacturing Management, Production Planning and Control, Optimization, Artificial Intelligence, Meta-heuristics, Scheduling, Web-based Systems, Services, and technologies, mainly for supporting Engineering and Production Management, Collaborative Networks, Decision-Making Models, Methods, and Systems, and Virtual and Distributed Enterprises for Industry 4.0. She has published more than 200 refereed scientific papers in international conferences and in international scientific books and journals, indexed in the Web of Science and/or in the Scopus databases. Coordinates R&D projects in the area of Production and Systems Engineering, namely, concerning the development of web-based platforms and decision support models, methods, and systems.

https://orcid.org/0000-0002-2299-1859
Email: leonilde@dps.uminho.pt

Goran Putnik received his B.S. and M.S. degrees in Production Engineering from the Faculty of Mechanical Engineering, University of Belgrade, and the DSc (Ph.D.) in mechanical engineering from the University of Belgrade, in 1993. He joined the Motor Industry Rakovica, Belgrade, Serbia, and 3 years later joined the Department of Production Engineering at Faculty of Mechanical Engineering, University of Belgrade as Assistant Researcher. In 1993, he joined the University of Minho, Portugal, where he is currently Full Professor in the Department of Production and Systems Engineering. His research and teaching interests include areas of production systems, automata, and formal theory. He is the author of 13 books, more than 250 articles, and coordinator of a number of international projects. Professor Putnik is member of the IFIP WG 5.5 – "COoperation infrastructure for Virtual Enterprises and electronic business – COVE", and of the International Academy for Production Engineering – CIRP, Paris.

https://orcid.org/0000-0003-3378-6866
Email: putnikgd@dps.uminho.pt

Dr. Vijaya Kumar Manupati is currently working as an Assistant Professor in the Operations and Supply Chain Division, Indian Institute of Management. He received his Ph.D. in the Department of Industrial and Systems Engineering from the Indian Institute of Technology Kharagpur. His research interests include intelligent manufacturing systems, agent/multi-agent/mobile-agent systems for distributed control, simulation, integration of process planning and scheduling, sustainable supply chain, and evolutionary algorithms. He has published more than 75 publications which include journals like *International Journal of Production Research, Computers, and Industrial Engineering, International Journal of Advanced Manufacturing Technology, Journal of Engineering, Journal of Measurements, International, Journal of Computer Integrated Manufacturing.*

https://orcid.org/0000-0002-3744-1315
Email: manupativijay@iimmumbai.ac.in